FIBRE REINFORCEMENT

Fibre reinforcement is receiving much attention due to the demand for high strength lightweight materials. The principles of fibre reinforcement are reviewed in a simplified, non-rigorous manner. The application of fibre reinforcement to metal, plastics and ceramic matrices and the benefits that accrue are described together with the range of reinforcements. Manufacturing processes are shown to achieve three objectives: to incorporate the fibres, to align them and to shape the product. Satisfactory processing techniques have been developed in all three matrix areas. The properties of fibre reinforced materials are reviewed. Extensive data exist in the metal and plastics matrix field and a summary is provided. In the ceramics field few data exist but it is clear that the brittleness of these materials is much reduced by fibre reinforcement. A wide range of applications of fibre reinforced materials is presented and the future for the technique is considered.

Other titles in this series

ME/1 TRIBOLOGY
E D Hondros

ME/2 ELECTRON BEAM WELDING
M J Fletcher

ME/3 VACUUM BRAZING
M J Fletcher

ME/4 STRESS CORROSION FAILURE
P Greenfield

ME/5 GLOW DISCHARGE MATERIAL PROCESSING
R A Dugdale

ME/6 DIAMOND GRINDING
J Burls

ME/7 ENGINEERING APPLICATIONS OF BERYLLIUM
P Greenfield

ME/8 WHISKERS
C C Evans

ME/9 CREEP OF METALS AT HIGH TEMPERATURES
P Greenfield

ME/10 CAVITATION
I S Pearsall

ME/11 MAGNESIUM
P Greenfield

ME/12 ZIRCONIUM IN NUCLEAR TECHNOLOGY
P Greenfield

ME/13 FRICTION WELDING
M J Fletcher

ME/14 METAL STRIP FROM POWDER
G M Sturgeon and R L S Taylor

ME/15 DEVELOPMENTS IN POWDER METALLURGY
R F Smart

M & B Technical Library

TL/ME/1 HYDROSTATIC EXTRUSION
J M Alexander and B Lengyel

TL/ME/2 THE HEAT PIPE
D Chisholm

TL/ME/3 PLASMA SPRAYING
R F Smart and J A Catherall

TL/ME/4 FATIGUE
J T Barnby

All published by Mills & Boon Limited

M & B Monograph ME/16

General Editor : J Gordon Cook, PhD, F R I C

Fibre Reinforcement

J A Catherall, BSc, AIM

Business Development Manager,
Foseco (F.S.) Ltd

Mills & Boon Limited

London

First published in Great Britain 1973
by Mills & Boon Limited, 17–19 Foley Street,
London, W1A 1DR.

ISBN 0. 263. 05119. 6

Made and printed in Great Britain
by Butler & Tanner Limited, Frome and London

CONTENTS

Preface

This monograph is intended as an introduction to the subject of fibre reinforcement and is aimed particularly at engineers and scientists who wish to gain first-hand knowledge of the field. The treatment is, of necessity, brief, but a more complete picture can readily be gained from the references in the bibliography. In the interests of lucidity, references have been omitted from the text.

The author wishes to thank the Directors of Associated Engineering Developments Limited for permission to use photographs and experimental data in this monograph. He is particularly indebted to those organisations and individuals who have granted permission to use their own original photographs.

1. Introduction

COMPOSITE MATERIALS

With the advance of technology, the need for the increasingly efficient utilisation of resources becomes ever more stringent. This applies nowhere more strongly than in the field of engineering materials. For many applications it is now becoming necessary to look to that class of materials known as composites. Composites, in the strictest sense, consist of any type of multiphase material, including such everyday materials as wood, concrete and steel. However, the term usually refers to multiphase materials in which the minor or dispersed phase is introduced deliberately by man rather than being allowed to develop fortuitously subject to the forces of nature. Thus, tungsten carbide–cobalt as used in cutting tools is a composite material, being manufactured by sintering and infiltrating. In some composites the dispersed phase is in the form of fibres, and such structures are known as fibre-reinforced materials. Thus, fibre reinforced materials are a particular type of composite material in which the dispersed phase is in fibrous form.

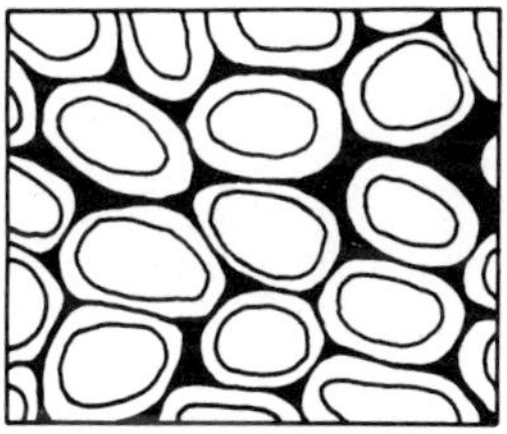

Fig. 1. Hollow cellulose fibres in a matrix of lignin–wood, a natural composite. Schematic diagram showing the structure magnified 1000 times.

The use of composites is by no means new. As already mentioned, wood is a composite; its structure can be seen in Fig. 1. Most of the structural materials of nature are composites, including for example, bones and teeth. The

Table 1. Representative ultimate tensile strength, S_u, and Young's Modulus, E, values of materials

MATERIAL	S_u (MN/m^2)	E (GN/m^2)	SPECIFIC GRAVITY (SG)	S_u/SG (MN/m^2)	E/SG (GN/m^2)
Silicon carbide	7,000	550	3·2	2,190	172
Silicon nitride	7,000	380	3·2	2,190	118
Graphite	23,000	1000	2·2	10,460	454
Alumina	15,000	410	4·0	3,750	103
Boron	3,000	440	3·3	909	134
Steel	3,000	210	7·8	385	26·6
Molybdenum	2,300	290	10·3	225	26·8
Aluminium	150	69	2·7	55·6	25·6
Wood	100	1·3	0·5	200	2·6
Nylon	70	0·3	1·1	63·6	0·28

legendary swords of history were composites in that the methods of manufacture introduced a favourably orien--ated dispersion.

The great dependence of the properties of composites on dispersed phase morphology is well demonstrated by the differences in behaviour of cross-ply and radial-ply tyres. The field we are particularly concerned with in this monograph, however, is that of high strength fibre reinforcement in which strong and stiff reinforcing fibres are incorporated in a weaker matrix.

BACKGROUND TO FIBRE REINFORCEMENT

The discovery of the extremely high stiffness and strength of single crystal filaments of materials, known as whiskers, was made by Herring and Galt in 1952. This discovery could be said to mark the beginning of interest in high strength fibre-reinforced composites. Whiskers are typically a few microns long and thus cannot be considered in isolation as engineering materials. (Fig. 2 shows a scanning electron photomicrograph of alumina whiskers.) It is necessary to incorporate them in a matrix so that the resulting composite can take advantage of the high load-bearing capability of the whiskers.

The strongest whiskers in terms of stiffness and ultimate tensile strength per unit of weight, and theoretically the strongest bulk materials, are the covalently bonded elements and compounds formed from the light elements of Groups II to VI of the Periodic Table. Thus silicon carbide, silicon nitride, graphite, alumina and boron are materials of high specific strength and stiffness. Table 1 shows representative values for these and other materials. In bulk form, these materials are extremely susceptible to the presence of minor defects, cracks or notches, and they behave in a brittle fashion. The high strength of whiskers is derived from their near-perfect structure, both internal and external.

Metallic and plastic materials do not have specific strength values as high as these ceramics; however, they are much more tolerant of surface defects and hence form the majority of engineering materials. Ceramics are brittle due to the inflexibility and high directionality of the covalent bond, whereas the bond in metals and plastics is flexible and much less directional, so that stress relaxation and redistribution can occur at notches.

In addition to whiskers, high strength metal wire, glass filaments and polycrystalline ceramic fibres all have applications in fibre-reinforced materials.

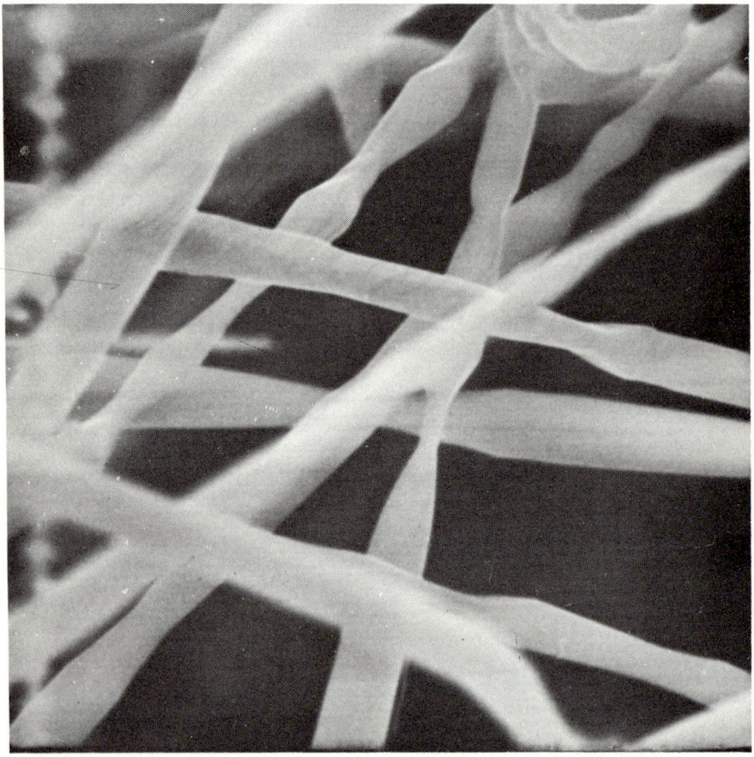

Fig. 2. Scanning electron micrograph of alumina whiskers ×5400.

STRENGTHENING MECHANISMS

Metals deform under stress due to the movement of internal defects known as dislocations (Fig. 3). The stress required to move a dislocation is much less than calculated for an en bloc movement of the material above the slip plane, so metals tend to deform easily.

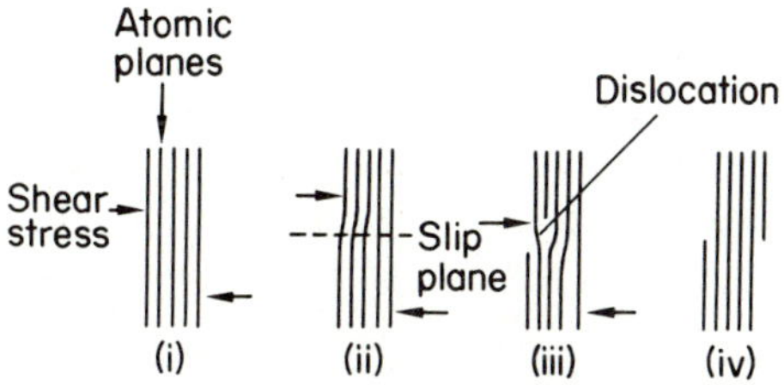

Fig. 3. Schematic diagram illustrating dislocation slip in metals.

The traditional approach to the strengthening of metals has been to impede the movement of dislocations. Thus atomic dispersions of, for example, carbon in steel or copper in aluminium are widely used to impede the motion of dislocations by pinning them in place or by making their passage less smooth. Similar effects are achieved by precipitation strengthening or work hardening. However, the mechanism of fibre reinforcement is not the result of such dislocation blocking. In a fibre reinforced material the low elastic modulus, low yield stress matrix strains or extends under the influence of stress. As it does so the fibres, bonded to the matrix as they are, extend by about the same amount, obeying the principle of combined action. The fibre is chosen for its high stiffness, so that at a given composite strain the load borne by the fibre is much higher than that borne by the matrix. This total load is shared by the matrix and the fibres, the latter carrying a more than proportional share.

In ductile polymeric materials, flow may not occur by the same mechanism as it does in metal matrices, but the result of fibre reinforcement is similar. In ceramics and glassy polymers, plastic flow does not occur to any sensible extent; fibre reinforcement of these materials is carried out for the quite different but equally important reason of inhibiting the propagation of cracks.

DESIGNING FOR FIBRE REINFORCEMENT

Effective fibre reinforcement depends upon the preferential loading-up of the high modulus fibres. This is achieved through matrix deformation. Conventional engineering is based on strains usually below 0·1% and to some degree a new engineering philosophy based on the acceptance of higher strains will need to be developed. This can result in problems, as pointed out by Professor J. E. Gordon. An aircraft with a strain of 1·6% in the wing spar booms would fly with its wing tips vertical (Fig. 4). At something over 3% the wing tips would meet over the pilot's head!

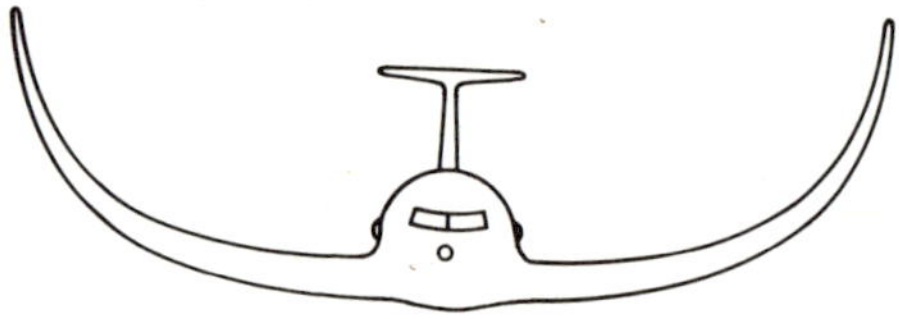

Fig. 4. Aeroplane flying with 1·6% strain in the wing spar booms.

Another problem could relate to the amount of elastic energy stored in a composite strained so that all the fibres are at their failure point. The high levels of strain energy stored in this way could result in a literally explosive failure of a component. These problems are, however, more apparent than real, since efficient use can be made of fibre reinforced composites at realistic strains. The anisotropy of their properties requires a careful design approach but should prevent few applications. A fracture mechanics approach to design, in which the stress level applicable to a given flaw size is calculated, is likely to be adopted with fibre-reinforced materials rather than the traditional approach of factor of safety, in the interests of efficient design. It is anticipated that a selective approach to fibre reinforcement, with only critical regions reinforced, will be adopted wherever possible for reasons of economy. The section through an automotive piston shown in Fig. 5 is the result of just such an approach.

Fig. 5. Automotive piston reinforced in the critical region with aluminium–stainless steel wire composite.

2. Principles of Fibre Reinforcement

Analysis of the mechanism of fibre reinforcement can be rendered complex in the absence of good fibre matrix bonding. If the interfacial bonding is good, however—which is usually the case with fibre-reinforced metals—then the following analysis is reasonably accurate.

ELASTIC DEFORMATION

The load carried by a composite material containing continuous fibres orientated undirectionally parallel to the load is simply the sum of the loads carried by the two components (Fig. 6):

$$L_c = L_m + L_f$$

where L refers to load and the subscripts c, m, and f refer to composite, matrix and fibre respectively. This may be

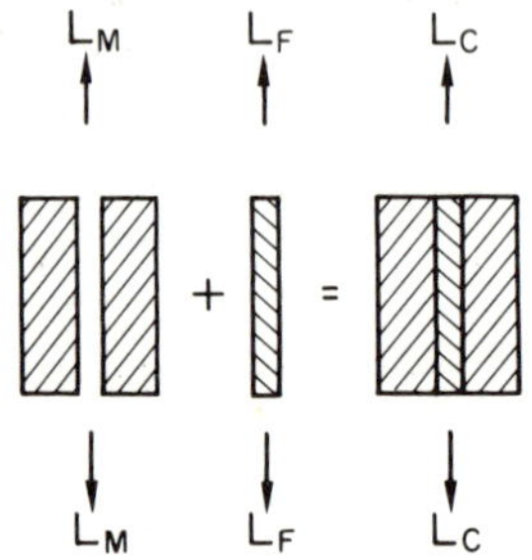

Fig. 6. $L_m + L_f = L_c$

expressed in terms of the products of the respective stresses (S) and areas (A):

$$S_c A_c = S_m A_m + S_f A_f$$

Since we are considering a material in which the cross-sectional area of each component is constant throughout the length, we may substitute volume terms for the area terms, and may then divide each side of the expression by the total volume, to give volume fractions of each component, represented by V.

Thus: $S_c V_c = S_m V_m + S_f V_f$
Now $V_m + V_f = V_c = 1$,
thus: $S_c = S_m V_m + S_f (1 - V_m)$

So the stress in the composite is equal to the sum of the stresses in the two components, weighted according to the volume fraction of each component.

If there is no slipping at the interface, the principle of combined action is said to be obeyed and e, the strain in each of the components, is the same.

$$e_c = e_m = e_f$$

Now Young's modulus, $E = S/e$.
Substituting Ee terms for S terms:

$$S_c = E_m e_c V_m (1 - V_m)$$

From this we see that for effective reinforcement, in which the majority of the load is carried by the fibres, it is necessary that we have the highest possible values of E_f and V_f since the ratio of load carried by the fibres to that carried by the matrix is given by

$$\frac{E_f e_c (1 - V_m)}{E_m e_c V_m} = \frac{E_f V_f}{E_m V_m}$$

To determine the Young's modulus of a fibre-reinforced material we substitute $E_c e_c$ for S_c as follows:

$$S_c = E_c e_c = E_m e_c V_m + E_f e_c (1 - V_m),$$

so that $E_c = E_m V_m + E_f (1 - V_m)$

Thus we see that the stiffness of a fibre-reinforced composite material in which the fibres are continuous and aligned parallel to the stress axis is simply the mean of the component values weighted according to their volume fraction. It follows that the Young's modulus of a fibre-reinforced composite is linearly related to the fibre volume fraction (Fig. 7).

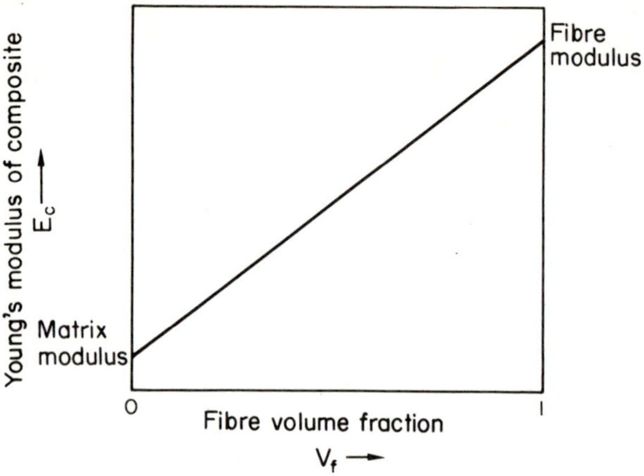

Fig. 7. Relationship between composite modulus and fibre volume fraction.

PLASTIC DEFORMATION

If the composite is loaded so that the yield stress of the matrix is exceeded then E_m is no longer the relevant term in the composite modulus expression. The slope of the matrix stress–strain curve under such conditions at a given value of fibre strain is represented by the differential term (dsm/dem)$_{ef}$. The value of this term is less than that of E_m and thus the composite modulus is lower under conditions of plastic matrix strain. Thus in the range of plastic matrix strain the modulus of the composite is represented by:

$$E_c = (dS_m/de_m)_{ef}\ V_m + E_f\ (1 - V_m)$$

On further straining a composite which contains ductile fibres, the point is reached at which the fibres themselves deform plastically. This results in a further deviation from the elastic line. In composites in which ductile fibres fail by necking, a hydrostatic state of stress arises in the matrix, in the absence of interfacial failure, which causes a deviation from the rule of mixtures relationship. Under such conditions, the portion of the matrix phase under hydrostatic tension can contribute a plastic "stiffness" up to three times the value in simple tension.

The stress–strain curve of an ideal composite is shown in Fig. 8.

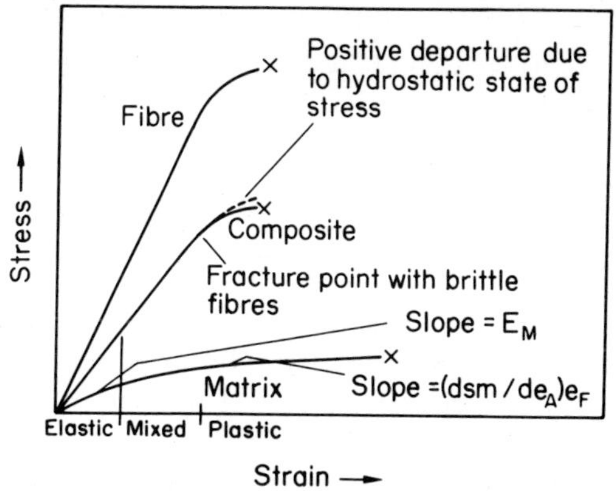

Fig. 8. Stress–strain behaviour of a composite containing 50% of fibres.

FAILURE

As we approach the failure condition a departure from the simple rule of mixture relationships can occur. This happens when there is less than a minimum volume fraction of reinforcement in a matrix which work hardens. In this case the failure strain of the fibres is reached while the net cross-section of the matrix is capable of supporting more load. The ultimate strength of the composite is then below that of the unreinforced matrix, since after the reinforcing fibres have failed the net area of matrix is reduced. The rule of mixtures is obeyed only if complete composite failure results when the fibres fail. Failure of the fibres causes failure of the composite when the ultimate stress of the composite exceeds that of the net cross section of matrix:

$$S_{cu} > S_{mu}\,(1 - V_f)$$

where the subscript u refers to ultimate values.
Now

$$S_{cu} = S_{fu}V_f + S_m{}^{uf}\,(1 - V_f)$$

where we use $S_m{}^{uf}$ to denote that the stress in the matrix is that prevailing at the ultimate strain of the fibres. Therefore for the rule of mixtures to be obeyed:

$$S_{fu}V_f + S_m{}^{uf}(1 - V_f) > S_{mu}(1 - V_f)$$

or
$$V_f(S_{fu} - S_m{}^{uf} + S_{mu}) > S_{mu} - S_m{}^{uf}$$

From which we define a minimum value of V_f, V_{min}, below which the rule of mixtures does not apply:

$$V_{min} = (S_{mu} - S_m{}^{uf})/(S_{fu} - S_m{}^{uf} + S_{mu})$$

Above this volume fraction the ultimate strength of the composite rises. At some volume fraction above V_{min} the strength of the composite is equal to that of the matrix alone. Below this critical value of V_f, which is called V_{crit}, no reinforcement occurs since the composite has a lower ultimate tensile strength than the matrix alone. Above V_{min} failure of the composite occurs immediately the fibres fail. The ultimate tensile strength of a composite with $V_f > V_{min}$ is thus:

$$S_{cu} = S_{fu}V_f + S_m{}^{uf}(1 - V_f)$$

For reinforcement, by definition, the composite must be stronger than the matrix alone:

$$S_{cu} > S_{mu}$$

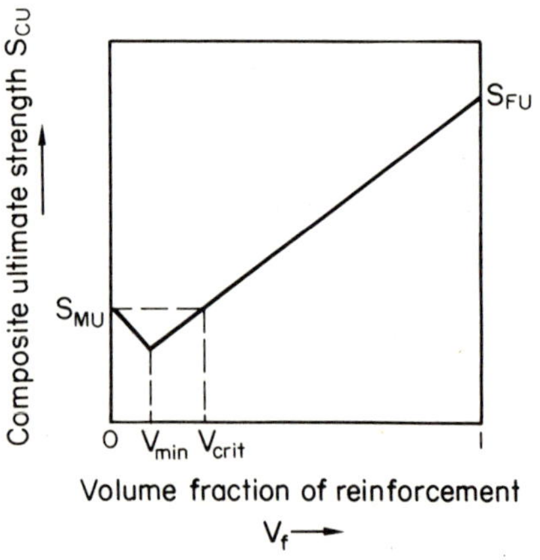

Fig. 9. Relationship between composite U.T.S. and fibre volume fraction.

Thus for reinforcement:

$$S_{fu}V_f + S_m^{uf}(1 - V_f) > S_{mu}$$

From this we derive the critical volume fraction below which no reinforcement occurs:

$$V_{crit} = (S_{mu} - S_m^{uf})/(S_{fu} - S_m^{uf})$$

The relationship between fibre volume fraction and composite strength is shown in Fig. 9. The expressions for V_{min} and V_{crit} assume that there is no change in the stress–strain behaviour of the matrix due to the presence of fibres and that all fibres are of equal strength and break in the same cross section.

NON-IDEAL BEHAVIOUR

The situation is more complex when the reinforcing fibres are not continuous. Since the fibres in a composite are loaded up via interfacial shear stresses, the end portions of discontinuous fibres are under lower stress than the central portion. A critical fibre length can be derived for a given diameter below which the fibre is not fully loaded by the interfacial shear stresses and reinforcement is inefficient (Fig. 10). For composites in which the fibres are markedly longer than the critical length, in which the area fraction is constant at any cross section, the stress–strain behaviour approximates to that described by the expressions for continuous fibre composites.

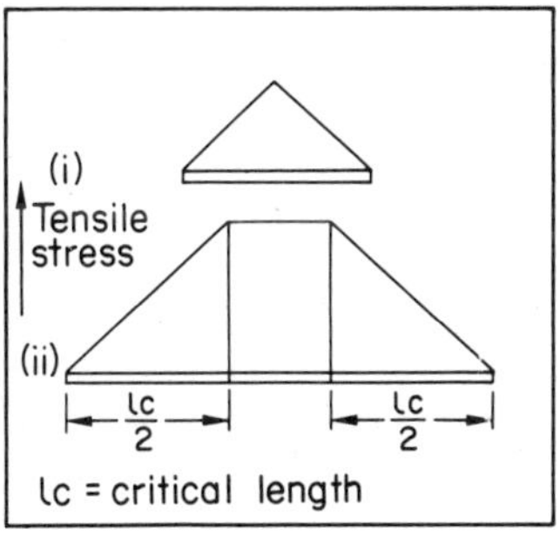

Fig. 10. Variation of tensile stress along a fibre of (i) less than and (ii) greater than critical length.

Values for a minimum and a critical volume fraction of fibres can be derived for discontinuous fibre composites in a similar fashion to those for continuous fibre composites. In general, however, the reinforcement obtained with discontinuous fibres is less than that obtained with continuous fibres.

In ceramic and plastic composites in which the matrix cannot deform plastically under increasing load, shear stresses in the matrix rise to a value at which either the matrix or the interface fails in shear. Load transfer continues under these conditions via frictional forces across the sheared region. Voids are considered to form at the ends of the fibres. High stress concentrations at the ends of the fibres can result in premature shear failure. It is found in composites of this type that transfer lengths are an order of magnitude higher than for well bonded metal composites.

We have considered so far idealised cases in which all fibres are of equal strength and orientated parallel to the stress axis. If parallel fibres are orientated at an angle to the stress axis then a rapid reduction in strength results with increasing angle, falling to a minimum at 45°, when failure of the matrix in shear parallel to the fibres is most favourable (Fig. 11). As the angle approaches 90° the stress rises.

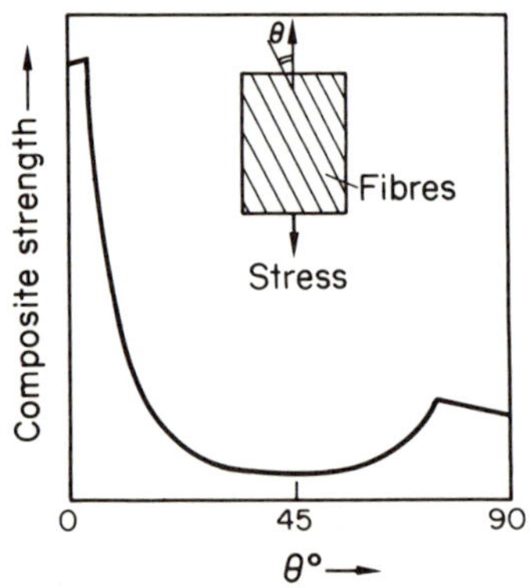

Fig. 11. Variation of composite strength with fibre orientation.

The transverse strength of a fibre-reinforced composite is dependent on the properties of the interface and the fibre volume fraction. Reasonably high transverse strength values can be obtained in metal matrix composites. However, for improved transverse properties it is possible to arrange the fibres in a random planar mat. Under these conditions the minimum strength in any direction should exceed one third of the unidirectional strength. Theory indicates that the strength of a three-dimensionally reinforced composite should be about one sixth of that of the unidirectional case.

Whilst metal wires of a given type can be assumed quite safely to have a strength of an approximately constant value, the same is by no means true for glass and ceramic fibres and whiskers. These generally show a considerable scatter in strength about the mean value. Whiskers, by their very nature, have a very variable strength since they vary considerably in size and flaw distribution. If it is assumed that failure occurs at the minimum diameter of a fibre, reasoning based on minimum-value statistics indicates that strength is dependent on length, short fibres being stronger than long ones. In a ductile matrix it is possible that long fibres will break at their weakest points to form shorter, stronger fibres, with the result that the composite strength exceeds that indicated by the measurement of individual fibre strengths.

COMPLEX STRESS BEHAVIOUR

Thus far, we have been discussing the behaviour of fibre reinforced composites under simple tensile loads. Under compressive loading it is considered that buckling of the fibres is a possible mode of failure. Two types of buckling have been proposed. In the extension mode the most prominent deformation is extension of the matrix between the out-of-phase fibres and in the shear mode shear of the matrix predominates (Fig. 12). For low fibre volume fraction the extension mode is the limiting factor in compressive strength, whereas at higher fibre contents the shear mode predominates. It is found in practice that the values indicated by theory for higher volume fractions of fibres are optimistic.

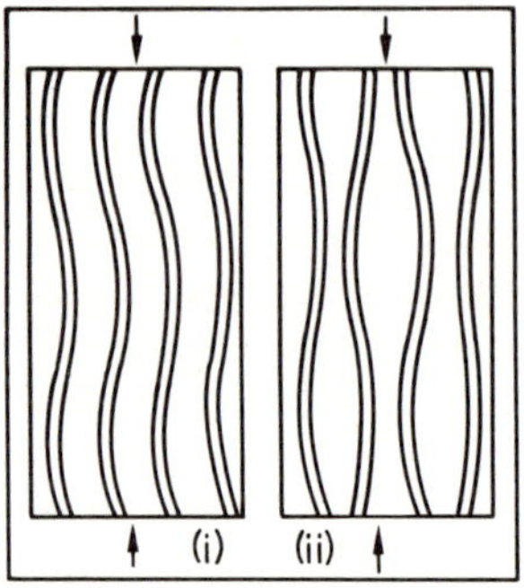

Fig. 12. Fibre buckling under compressive loading, (i) shear mode, (ii) extension mode.

Under conditions of repeated stressing, when fatigue might occur, a number of important effects must be noted in fibre reinforced materials. These materials have a high damping capacity and, on repeated stressing, a large quantity of heat can be evolved which is especially troublesome in the case of reinforced plastics. Under fatigue loading the stress in the matrix of a composite is lower than that which would occur in the unreinforced matrix, since the fibres carry a more than proportional share of the load. Thus, the initiation of cracks can be expected to be delayed and their subsequent propagation rate is likely to be much reduced. In addition, the interfaces may act to deflect or stop cracks by opening up to produce a void. The effect of reinforcement on fatigue life should thus be very significant.

Creep, high temperature deformation under constant load, is very sensitive to matrix stress and this is likewise expected to be an area in which fibre-reinforcement has much to offer.

The impact strength or toughness of fibre-reinforced materials is a property that can arise from the ductility of the matrix in its ability to blunt cracks or notches by flow, or from delamination in which the fibre-matrix interface fails, resulting in a material very much like bamboo with a large work of fracture. When a crack is present in a fibre-reinforced composite which is stressed parallel to the fibres, a transverse tensile component of stress exists ahead of the crack tip. It has been proposed that this component of stress will exert a major effect on toughness by causing the opening-up of weak interfaces as shown in Fig. 13. The crack is thus deflected along the interface.

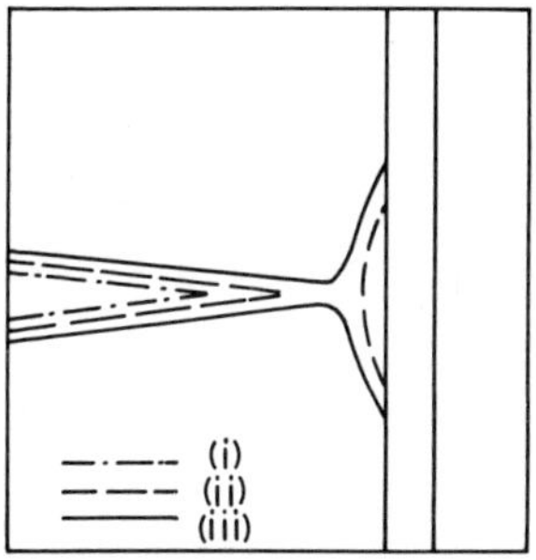

Fig. 13. Crack approaching fibre (i), opening-up the interface (ii), and merging with interface (iii).

A third approach to toughness has been proposed in which either the matrix or the interface is weak in shear. This results in large transfer lengths and an inability to concentrate the stress at a fractured fibre. Under such conditions a propagating crack can only be produced at the highest stresses and the mode of failure, in which the fibres are pulled out of the matrix, results in a large work of fracture if the shear stress is maintained at the interface during pull-out.

3. Materials

Most interest in the materials aspects of fibre reinforcement centres around the reinforcement, as the agents of the mechanism. However, it is instructive to consider the types of matrices available in order to understand how they benefit from fibre reinforcement. Fig. 14 summarises the position.

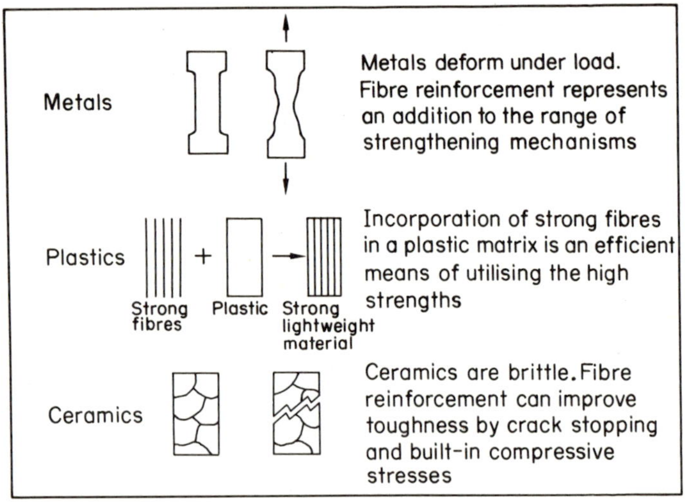

Fig. 14. The benefits of fibre reinforcement as applied to the range of matrices.

MATRIX MATERIALS

Metals and their alloys represent the class of materials most widely used in engineering. They are relatively easy to shape by one method or another, strong and, even more important, they are tough. The last property means that they will not shatter under the influence of impact. They avoid such failure by their ability to deform plastically. Gross plastic deformation is usually a problem, however, and for the most efficient use of a material we may require the highest possible yield stress. Reinforcement with stiff fibres raises yield stress and thus allows of more efficient design. The matrix stress is reduced and a fracture mechanics approach to design will indicate increased load capacity.

Metals are subject to two common types of failure: creep and fatigue. Creep occurs at high temperatures and is very sensitive to matrix stress. A reduction in stress by the incorporation of strong fibres can be expected to exert a major effect on creep life. Fatigue is by far the most common cause of failure in metals. It is caused by the initiation and propagation of cracks. The presence of strong fibres can inhibit the initiation and propagation of such cracks and eventually stop them altogether.

In these ways, fibre reinforcement can extend the limits of metallurgical technology. A new approach to materials engineering is possible in which less compromise is necessary. Matrices can be chosen for their surface properties, or a weak alloy with some desirable bulk property such as electrical conductivity can be used in reinforced form. For example, pure metals are not generally susceptible to stress corrosion, whereas their alloys are. Weak pure metals could be used in a stress corrosion situation reinforced to fulfil mechanical property requirements.

It is instructive to consider plastic matrices from the point of view of fibre utilisation, since incorporation of fibres in plastic matrices can be regarded as the most powerful way of utilising the high strengths available. In contrast, fibre reinforcement of metals can be regarded as merely an extension of the currently available range of strengthening mechanisms. Plastics are the newest range of materials and they are available in a multitude of forms. However, in terms of their application in general engineering they tend to exhibit low elastic modulus and quite often low strength as compared with metals. They are extremely attractive, however, from the point of view of density. Their low elastic modulus, in general an order of magnitude below that of aluminium, enables them to derive very substantial benefit from stiff fibres. The incorporation of fibre is generally easy and fibre loadings up to about 70% are realisable. The resulting products are extremely stiff and strong with low density.

Ceramics and glasses have many properties of interest to engineers but their inherent brittleness has prevented their widespread use in stressed applications. Fibre reinforcement of these materials has not been widely studied, but two advantages are offered by the approach. Fibre reinforcement can provide weak interfaces through which cracks cannot easily run. Probably more important, a compressive stress can be built into the material by the incorporation of fibres under tension. The stress concentration which occurs at the root of quite large flaws could thus be cancelled out. Such an approach could yield an extremely tough ceramic or glass which would enable its other interesting properties to be utilised.

REINFORCEMENTS

Fibres for reinforcing purposes fall into four distinct groups: (*a*) metal wires, (*b*) glass fibres, (*c*) ceramic fibres, and (*d*) single-crystal whiskers.

(*a*) *Metal Wires*

Metal wires represent one of the cheapest and most readily available forms of reinforcement. Table 2 lists a range of metal wires with their relevant properties. The $\frac{2}{3} \times$ melting point figure is included as a measure of temperature capability and of, to some degree, stability.

Ferrous and refractory metal wires have been widely used in studies of fibre reinforcement and for many purposes they are extremely useful. However, for applications requiring low density they have disadvantages. Hard drawn carbon steel wire, piano wire, is an ideal material where low cost is a major factor. However, the low tempering resistance of this material limits its application. Stainless

Table 2. Metal wires: summary of properties of relevance to fibre reinforcement

	RELATIVE COST (*)	YOUNG'S MODULUS (GN/m^2)	ULTIMATE TENSILE STRENGTH (MN/m^2)	SPECIFIC GRAVITY	$\frac{2}{3}$ × MELTING POINT (K)	REMARKS
Piano wire	1	210	4000	7·8	1000	Low tempering resistance
Stainless steel	1	200	4000	7·9	1000	
Molybdenum	10	290	2300	10·3	1900	
Tungsten	20	350	4600	19·3	2500	
Beryllium	500	300	1200	1·8	1000	Not widely available

Costs and properties are highly sensitive to wire diameter. Comparable representative figures are given.
* Per unit of weight.

steel wire with its improved tempering resistance is used in applications where temperatures up to 800K can be expected. Stainless steel can be incorporated in aluminium by hot pressing techniques without a significant fall in yield strength. The Young's modulus of a material is not affected by tempering, so piano wire is suitable for applications requiring stiffness rather than ultimate strength.

Refractory metal wires, which have an extremely high Young's modulus, have been widely used in fibre reinforcement; availability, especially in the case of tungsten, which is used in lamp filaments, is also an important factor. Tungsten wires in conjunction with copper matrices have contributed to the understanding of fibre reinforcement in a very significant way since they are favoured in model systems due to their immiscibility.

Superalloys, for use in creep and fatigue environments at temperatures around 1300K, have been reinforced by the incorporation of refractory metal wires. At these temperatures, when all conventional engineering materials are approaching or well above maximum service temperature, refractory metals are, relatively, barely warm. In a properly designed composite of this type the refractory metal wires will be behaving elastically, restraining matrix flow.

Tungsten is also used in the fibre reinforcement field as a substrate for fibres manufactured by vapour deposition; for example, boron fibre.

Beryllium is a particularly interesting metal reinforcement wire. The exploitation of beryllium on a major scale has been hindered by problems of toxicity and brittleness, but both are overcome if beryllium is used in wire form for fibre reinforcement. Beryllium has a density only two thirds of that of aluminium and a Young's modulus almost half as high again as that of steel. Its melting point of 1557K indicates a temperature capability in excess of 1000K. These properties serve it well in fibre reinforcement applications.

The drawing of fine metal filaments suitable for use in fibre reinforcement is an expensive process. Alternative routes for the manufacture of metal filaments have been investigated. These involve manufacture from foil (Figs. 15 and 16) and turning from bulk. Fibres produced in this way are relatively cheap and have elastic properties similar to those of hard drawn materials. It is probable that swarf of the right kind could be used as a reinforcing filler, but the high aspect ratio required for efficient fibre reinforcement would seem to militate against the use of by-product swarf.

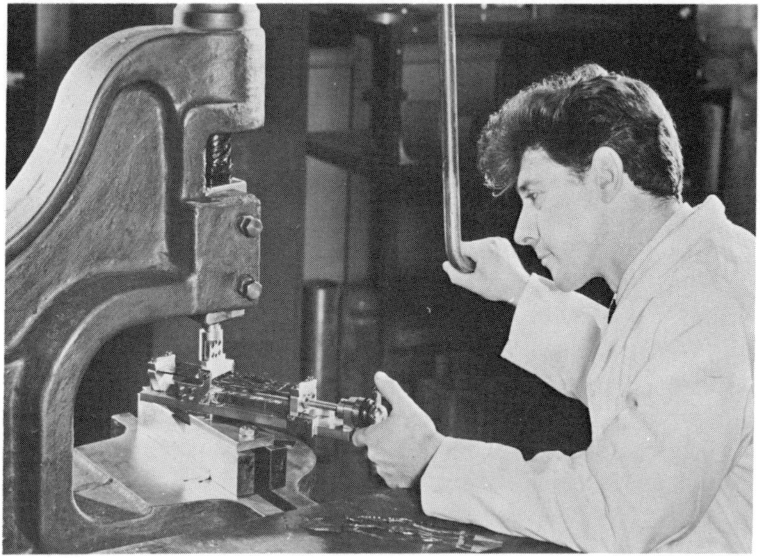

Fig. 15. The manufacture of metal filaments from foil.

Two other techniques are worthy of mention, both of which involve the encapsulation of coarse reinforcing fibres in a softer matrix. In *bundle drawing*, a number of coarse filaments are encapsulated and drawn through a series of dies. The soft matrix is leached away after drawing. Metal and glass filaments have been manufactured in this way. In the *Taylor process*, a metal-filled glass tube is drawn to fine diameter. Both processes are capable of producing fine high-strength wires.

Fig. 16. Tool used in the manufacture of filaments from metal foil.

Ferrous reinforcements find application in the reinforcement of low melting point materials, light alloys and copper alloys in which the weight penalty is not a significant factor and cost of reinforcement is. Since cost of reinforcement is likely to be overshadowed by process costs, it is unlikely that ferrous metals will find application in fibre reinforcement; more suitable materials are usually available. For example, for the highest temperature applications, refractory metals will find a use, in competition with ceramic fibres. At lower temperatures, say below 1000K, beryllium is superior to ferrous materials in all respects but cost. If fibre reinforcement is ever required for a high volume low cost product, then a ferrous filament produced direct from bulk stock may find application. Such a use would require a cheap and efficient incorporation technique.

(*b*) *Glass Fibres*

Glass filaments represent a low cost, readily available

reinforcement which is to the plastic matrix field what metal wires are to the metal matrix field. Very high ultimate tensile strengths can be achieved from glass fibres when their surface is perfect. 3500 MN/m^2 is easily attained with E-glass. The modulus of glasses falls in the range 50–110 GN/m^2. The elastic modulus of plastics is usually substantially bellow 7 GN/m^2, so we have an order of magnitude difference in modulus. The density of glass is comparable with that of plastics and the chemical behaviour of plastic matrices is not adversely affected by the presence of glass fibres.

Fine high strength glass fibres are manufactured by extruding molten glass through an orifice, typically about 2·5 mm in diameter, and then rapidly pulling to draw it to a fine diameter, typically 0·01 mm. In practice a large number of orifices are used and the individual fibres are combined in one strand. Prior to this a protective coat is applied to the fibres to protect their surface, since glass fibres are highly abrasive to one another.

Strand and roving are the cheapest forms in which glass fibre is available. Strand is defined above; roving is a bundle of parallel strands. In these forms glass fibres are used in filament winding and prepreg techniques. Strand or roving may be chopped to produce chopped fibre. Non-woven mats of glass fibre are used in which the individual fibres are held together by a resin binder or by continuous strands distributed through the chopped fibre. For the highest volume fraction of reinforcement woven fabrics are used. These vary in complexity and cost, from woven roving to multiple-ply twisted yarn glass-fibre cloth.

As a result of the very rapid cooling during fibre formation, the structure of glass fibres is similar to that of bulk glass at elevated temperature, with increased interatomic spacing and modified atomic arrangement as compared with bulk glass at room temperature. These changes result in fibres having reduced density and elastic modulus; the properties of glass fibres cannot be equated with those of bulk glass. These effects are stable up to about 373K, above which the reversion to the normal structure occurs in a process known as thermal compaction.

The most widely used type of glass for fibre manufacture is E-glass, a lime–alumina–borosilicate glass developed for high electrical resistance, which has good fibre-forming properties. The compositions of the fibre-forming glasses are presented in Table 3 and their properties in fibre form in Table 4. Common soda lime glass (A-glass), as used in bottle and window manufacture, is used in certain reinforcing applications. C-glass has superior chemical resistance to E-glass. D-glass was developed for its low dielectric constant for radome applications and is characterised by a low specific gravity, 2·16. The addition of beryllia increases the modulus of glass fibres quite substantially; M-glass containing 8% BeO has an as-drawn modulus of $110GN/m^2$. S-glass was developed to provide a glass fibre with higher strength and elastic modulus than E-glass, together with improved elevated temperature strength. Fused silica is a simple glass which does not exhibit thermal compaction and is characterised by low density, low thermal expansion coefficient and high strength.

Glasses have extremely high tensile strength and it was a study of this high strength which led Griffith to expound the theory of the effect of flaw size on strength. This approach has been widely applied to the fracture of brittle solids and forms the basis of fracture mechanics. According to the Griffith theory a brittle solid contains a distribution of surface flaws, and fracture will occur at the largest flaw when the strain energy contained in the stressed body is equal to the surface energy of the fracture surfaces (Fig. 17). For a crack of length C which is atomically sharp, the lowest stress, S, for its propagation is given by the Griffith formula:

$$S = \left(\frac{2E\gamma}{\pi C}\right)^{\frac{1}{2}}$$

where E is Young's modulus, and γ is surface energy. For

Table 3. Typical compositions of glass fibres

	PER CENT						
	SiO_2	Al_2O_3	CaO	MgO	B_2O_3	Na_2O	OTHERS
E-glass	54	14	19	2	10	1	K_2O, TiO_2, Fe_2O_3, F_2
A-glass	72	1	10	2		14	SO_3 1%
C-glass	65	4	14	3	5·5	8	K_2O 0·5%
M-glass	54		13	9			BeO 8%, TiO_2 8%, CeO_2, Li_2O, ZrO_2, Fe_2O_3
S-glass	65	25		10			

Table 4. Glass fibres: summary of properties of relevance to fibre reinforcement

	RELATIVE COST (*)	YOUNG'S MODULUS (GN/m^2)	ULTIMATE TENSILE STRENGTH (MN/m^2)	SPECIFIC GRAVITY	STRAIN POINT (K)	REMARKS
E-glass	0·2	72	3400	2·54	889	Most readily available fibre
A-glass	0·2	70	3200	2·50	800	Common soda-lime glass
C-glass	1	69	3100	2·49	825	Corrosion-resistant
D-glass	3	52	2400	2·16	748	High dielectric constant glass
M-glass	2	110	3400	2·89	800	High modulus
S-glass	0·5	85	4800	2·48	1033	Developed to give higher strength than E-glass
Fused SiO_2	5	73	7000	2·20	1343	

* Steel wire = 1 per unit of weight
Strain point is a measure of temperature capability, being the temperature at which the viscosity is $10^{13.5}$ N s/m^2.
Costs and properties are dependent on fibre diameter; comparable representative figures are given.

E-glass the strength of E/20 commonly achieved corresponds to a maximum crack length of only about 0·0015 μm. This indicates that the flaws are only of atomic dimensions or that they are not atomically sharp. Internal flaws in glass fibres do not appear to affect strength significantly.

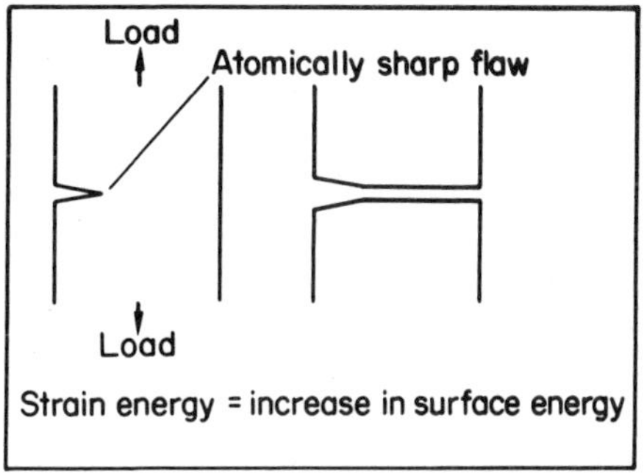

Fig. 17. Representation of Griffith theory.

Moisture is known to have a deleterious effect on the strength of glass fibre. Physico-chemical adsorption produces local areas of corrosion and this reduces strength probably due to a combination of stress corrosion and a reduction in effective surface energy. Static fatigue, under which glass will fail after an extended time under a static load, occurs on a similar basis due to the growth of pre-existing flaws.

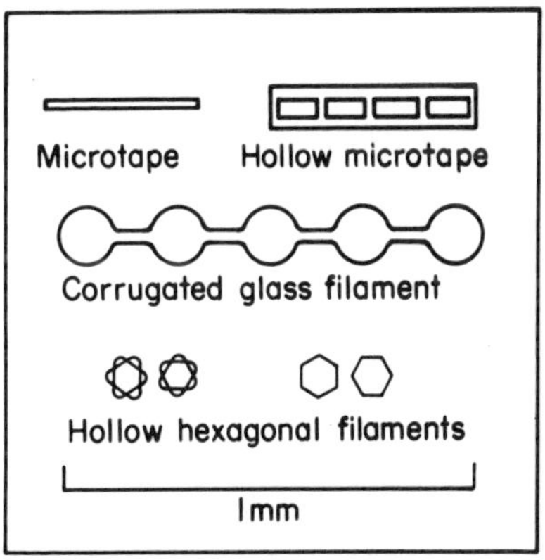

Fig. 18. Types of shaped glass fibre.

A solid cylinder has shortcomings when it is desired to achieve the highest packing fractions. For this reason, interest has been shown in the development of shaped glass fibres. In addition, fibres can be made hollow to achieve low density coupled with the highest flexual stiffness. Greater control over the anisotropy of glass reinforced plastics can be achieved by using fibres of complex cross section. Fig. 18 shows a selection of shaped glass fibres.

(c) Ceramic Fibres

To achieve the highest possible strength, stiffness and temperature capability at low density it is necessary to turn to the brittle refractory covalent and ionic crystals of complex structure and to intermetallic compounds. Unlike metals, these materials do not suffer from significant irreversible flow at temperatures between 0·5 and 0·7 T_m, where T_m is the melting point on the absolute scale. Only above the upper figure does the thermally activated movement of dislocations cause large amounts of irreversible deformation in these materials.

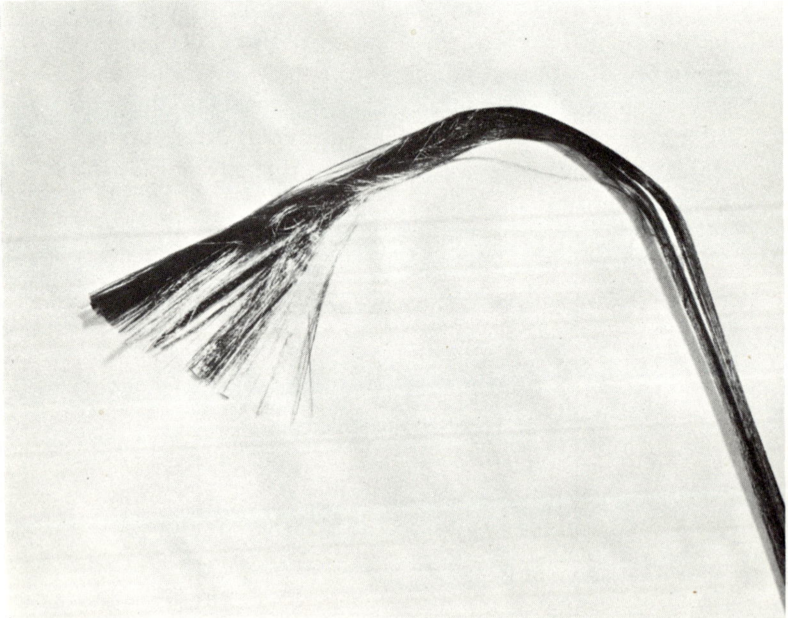

Fig. 19. High strength carbon fibre.

Carbon fibre is an example of a polycrystalline ceramic fibre. Carbon fibres have been produced for many years, but early fibres had low modulus values. In the early 1960s, graphite whiskers were produced with a Young's modulus approaching 700 GN/m^2, which indicated the potential of this element. Graphite is highly anisotropic with layers of hexagonal-structure plates piled one on top of the other. The highly directional covalent bond of the basal hexagonal plane results in a Young's modulus parallel to the plates of 1000 GN/m^2. The Young's modulus perpendicular to the plates is governed by weak van der Waal's forces and is only about 35 GN/m^2. Thus, a polycrystalline fibre in which crystallites are highly orientated parallel to the fibre axis should exhibit a high modulus. The currently available carbon fibres (Fig. 19) achieve a high degree of orientation during a high temperature treatment in which tension is applied to the fibre. The first step in the manufacture of carbon fibre is the extrusion of polyacrylonitrile through a spinnerette containing 10,000 holes. The textile fibre thus produced is heated in air at 500K whilst being wound on a frame under tension, to prevent shrinkage. A ladder polymer is formed, which subsequently oxidises. The temperature is raised to 1300K and carbonisation occurs in an inert atmosphere. Hydrogen cyanide, ammonia and water are evolved, producing fibres with a carbon content of 95%. Graphitisation is performed at temperatures between 1800K and 2800K. The properties of the product are determined by the temperature and tension during graphitisation. The fibres with the highest tensile strength are produced by low temperature graphitisation, those with the highest modulus by high temperature graphitisation.

Table 5 presents the properties of the two types of fibre. Apart from the properties of particular significance in fibre reinforcement, graphite fibre has other interesting and useful properties. The coefficient of friction perpendicular to the fibre axis is very low, giving rise to interesting tribological possibilities. The fibre has a negative coefficient of expansion, so that the matrix is compressed as the temperature rises. Whereas the matrix becomes

Table 5. Polycrystalline ceramic fibres: summary of properties of relevance to fibre reinforcement

	RELATIVE COST (*)	YOUNG'S MODULUS (GN/m²)	ULTIMATE TENSILE STRENGTH (MN/m²)	SPECIFIC GRAVITY	$\frac{2}{3}$ × MELTING POINT (K)	REMARKS
I C-fibre HM	15	410	1700	2·0	3920^s	High modulus fibre
II C-fibre HT	15	280	2700	2·0	3920^s	High tensile fibre
B-fibre	200	380	2700	2·6	1700	Hardness > 9 mohs scale
Alumina	200	170	2000	3·1	1550	
Zirconia	200	340	2000	4·8	1950	
Asbestos	0·01	190	6000	2·5	773^d	
Mica	0·01	320	2700	2·7	673^d	Platelets

* Steel wire = 1 per unit of weight

^{s}—sublimes.

^{d}—dehydrates.

Where size variations can occur prices and properties are dependent on diameter; comparable representative figures are given.

weaker in general as the temperature rises, the strength of graphite fibre rises marginally and continues rising up to 3200K; above this temperature the graphite sublimes.

The development of carbon fibre has been overshadowed in the U.S.A. by the strategic development of boron fibre. Boron was produced in excess of the amounts needed for nuclear engineering applications, and a survey was performed to identify potential markets; as a result, the attraction of boron as a reinforcing fibre was established.

Boron cannot be formed by conventional means and it has been necessary to develop vapour deposition techniques for the manufacture of boron fibre. Two production routes are possible. In the first, a boron halide is reduced by hydrogen and the boron is deposited on a heated 0·01 mm diameter tungsten wire maintained at about 1500K. The chemical reaction may be represented by the equation.

$$2BX_3 + 3H_2 \rightarrow 2B + 6HX,$$

where X represents Cl, Br, I.

The alternative production route utilises the thermal decomposition of boron hydrides:

$$2B_x H_y \rightarrow 2xB + yH_2$$

The latter route is advantageous in that the process can be performed at lower temperatures, but fibres produced by this route tend to be weaker, poorly bonded to the substrate and less dense. A microcrystalline deposit is formed on the tungsten substrate. Boron diffuses to some extent into the tungsten, forming tungsten borides.

The properties of boron fibre are presented in Table 5. Of particular interest are the very high strength, the high modulus and the low density. The hardness is significant in that the individual fibres are not so susceptible to surface damage as are, for example, glass fibres. The incorporation of boron fibres into matrices other than plastics is considerably easier than for carbon fibre, due to the increased stability of the former. Boron fibre will probably remain an order of magnitude more expensive than carbon fibre, which is itself struggling against unfavourable economics. In common with silicon carbide and other fibres manufactured by a vapour deposition route, boron fibre has a larger diameter, 0·1 mm, than other reinforcements which vary between 1μm and 25μm. Boron fibre is suitable for use up to temperatures in excess of 1300K.

Other types of polycrystalline ceramic fibres have been the subject of considerable research, particularly in the U.S.A. A wide range of fibres has been made by dispersion of the powdered fibre material in a plastic binder and extrusion of the mixture through an orifice. A sol-gel route has been developed in the U.K. at A.E.R.E., Harwell. Firing of the fibre produced drives off residuals and densifies the material. Alumina, zirconia and carbon fibres have all been manufactured by this route. Strengths approaching 6900 MN/m^2 have been recorded for the finest filaments.

Fig. 20. Asbestos, a cheap fibrous reinforcement.

Cost and availability are, as indicated above, problem areas in the exploitation of high strength reinforcements. Two reinforcements, asbestos (Fig. 20) and mica, are cheaper than glass fibre, and they can be dug from the earth. They have properties which are in many respects superior to those of glass fibres. Mica is plate-like in form, and it cannot strictly be regarded as a fibre reinforcement. However, platelet reinforcement is in reality an extension of fibre reinforcement to two dimensions, with the added advantage that much higher packing fractions are attainable.

Both asbestos and mica can operate up to the limit of currently available technological plastics without any loss of strength. Asbestos is, of course, widely used as a filler in friction materials. The strength and stiffness of similar material combinations will be fully utilised in applications in which loading is not mainly compressive.

(d) Whiskers

Whiskers represent the ultimate in reinforcement in that they exhibit properties well outside the range of those of the other classes discussed. Whiskers are single-crystal acicular growths, typically 5 mm long and 0·01 mm thick, with near-perfect atomic arrangements; this is important in materials which deform by the movement of dislocations, and they exhibit near-ideal strengths when their surface is undamaged. A wide range of metal and ceramic whiskers has been developed and studied; ceramic whiskers are to be preferred on grounds of strength, refractoriness and stability. The properties of a selection of whiskers are presented in Table 6.

The growing of whiskers is a slow process and the harvesting and handling of the product is extremely difficult. These factors account in large measure for the high cost of whiskers. Techniques of manufacturing and handling short-length fibres for reinforcement purposes have been brought to a high level of sophistication in the U.K. by the Explosives Research and Development Establishment. Silicon carbide and silicon nitride whiskers are manu-

Table 6. Single crystal whiskers: summary of properties of relevance to fibre reinforcement

	RELATIVE COST (*)	YOUNG'S MODULUS (GN/m^2)	ULTIMATE TENSILE STRENGTH (MN/m^2)	SPECIFIC GRAVITY	$\frac{2}{3}$ × MELTING POINT (K)	REMARKS
Silicon carbide	5	480	7,700	3·21	2000	Large-scale batch production
Silicon nitride	10	380	7,700	3·18	1900^{s}	Large-scale batch production
Alumina	250	420	15,000	3·96	1550	
Graphite	200	690	20,000	2·20	3920^{s}	
Iron	200	200	8,000	7·83	1200	

* Steel wire = 1 per unit of weight.
^{s}—sublimes.
Whiskers vary considerably in size and growth orientation and properties and prices vary accordingly; comparable representative values are given for good quality whiskers.

factured in batch processes in which a powdered mixture of silicon and silica is heated to 1700–1800K in the presence of gaseous reactants. The whiskers nucleate and grow on vertical rods. The resulting fibres are separated in a liquidiser and graded in an elutriator.

Random mats and aligned yarns are produced. Figs. 21 and 22 show respectively an optical micrograph and a stereoscan photograph of good quality β-Silicon carbide whiskers as grown.

Fig. 21. Optical micrograph of good quality -SiC whiskers as grown ×175. (Courtesy Dr. C. C. Evans, E.R.D.E.)

The most widely studied whiskers are those of alumina or sapphire. These are produced normally by the reaction of aluminium with water or aluminium oxide with hydrogen at 1600–1800K.

$$2Al + H_20 \longrightarrow Al_20 + H_2$$
$$Al_20_3 + H_2 \longrightarrow Al_20 + 2H_20$$

Fig. 22. Stereoscan photograph of good quality -SiC whiskers as grown ×250. (Courtesy Dr. C. C. Evans, E.R.D.E.)

The suboxide decomposes to aluminium oxide and aluminium in a cool zone of the furnace:

$$3Al_20 \longrightarrow Al_20_3 + 4Al$$

When the decomposition rate is low, whiskers of alumina are formed. Alternatively, an aluminium-containing intermetallic compound can be used as starting material. An interesting route has been developed by Grimshaw and Peacey in which reaction of bauxite with aluminium trichloride and carbon produces a sub-halide which subsequently decomposes in a cool zone to yield the alumina whiskers shown in Fig. 2.

Graphite whiskers are grown by a sublimation process in which an arc is struck between graphite electrodes. Scroll-like graphite whiskers form in an agglomerated mass.

Intermetallic constituents of alloys are extremely strong single-crystal growths. It is possible to grow them *in situ* by directional solidification of a eutectic as described below. Such whiskers can have strengths approaching 7000 MN/m^2.

High strengths can be developed in bulk materials, but the important feature about whiskers is that they are strong as a rule rather than only in exceptional cases. As there are millions of whiskers in a composite, the failure of a few can be borne by the remainder. Surface flaws therefore are less troublesome than they would be in a high strength bulk solid. High modulus is probably the most important property of a whisker, in that it enables high stresses to be developed at usable strains. Modulus varies in crystals with crystallographic direction, and whiskers—being single crystals—can exhibit extremely high modulus values. Plastic flow is also orientation-dependent, since for such irreversible deformation to occur a slip system must be aligned close to a direction of high shear stress. For example, the (1120) slip-plane in sapphire is orientated in such a way in C-axis whiskers that no resolved shear stress is available along the slip plane to produce plastic flow. Thus, whiskers can exhibit usable strengths well in excess of those of other reinforcements. The refractory whiskers hold most promise, silicon carbide and sapphire being two of the best. They are characterised by high usable strength, low density, inertness, high melting point and extreme resistance to plastic deformation at temperatures up to a very high proportion of their melting point.

4. Composite Fabrication Processes

Composite fabrication processes involve not only the incorporation of fibres in a matrix but also the alignment of the fibres and the production of the finished component shape. To meet these ends a wide range of highly sophisticated techniques has been developed.

POLYMER MATRIX COMPOSITES

Techniques for the manufacture of glass reinforced plastics have been developed over the years and are readily applicable to the newer types of reinforcement.

In the field of continuous fibre reinforcement low pressure moulding techniques are widely used in which hand lay-up in a shaped half-mould is followed by coating with liquid resin and curing with or without the application of pressure to give in excess of 40% volume fraction of reinforcement. Matched die moulding is used for both continuous and discontinuous reinforcements to give up to 60% volume fraction. Extrusion or pultrusion is performed by passing continuous yarn or roving through a resin bath and squeezing it through a die. Up to 70% volume fraction of reinforcement can be achieved in this way. Filament winding is the process which has found widest application in the manufacture of high strength structures. Roving or flat tape is passed through a resin bath and wound onto a mandrel under tension and cured. Pre-impregnated roving or tape, prepreg, may be used to eliminate the resin bath and in this case a heated roller is used to impart a degree of self-adhesion. Up to 80% fibre volume fraction can be achieved.

Discontinuous fibres in the form of chopped strand mat are used in a variety of processes from bucket and brush to sophisticated gun techniques for the reinforcement of resin matrices. Premix or dough-moulding compounds also utilise discontinuous fibres in addition to a wide range of mineral fillers. These are readily shaped and cured by the application of heat. Whiskers are incorporated into polymer matrices by infiltration of a mat or bundle with a low viscosity resin or with a solution of the resin in a solvent carrier.

Fig. 23. Fibre reinforced PTFE coating on steel produced by electrophoresis and sintering ×150.

Due to the forming processes used in the thermoplastic field in which the materials is simply heated, formed and allowed to cool, discontinuous reinforcements are more readily used than the continuous form. The basic raw material is usually pellets containing the reinforcement. These are processed in the normal way to produce the product. The pellets themselves are made by hot-impregnating a continuous form of the fibre and then chopping. Electrophoresis has been applied to the manufacture of P.T.F.E. composites and this approach has potential in the fibre-reinforcement field (Fig. 23). In a newer technique, reinforced thermoplastic moulding compounds are produced in a one step interfacial polymerisation process. The polymer is prepared from two fast-reacting intermediates near an immiscible solid-liquid interface. A polymer rope forms at the interface and is drawn off. Short fibres are held in suspension and enter the polymer with minimal damage.

METAL MATRIX COMPOSITES

The most significant area in the high strength composite field, in terms of potential if not of immediate application, is that involving metal matrices. In contrast to the high strength reinforced plastics field, no well established technology was available and many novel techniques have evolved. These can be divided conveniently into three approaches according to whether incorporation is performed in the presence of a solid phase, a liquid phase, or atomic species.

(*a*) *Solid State Incorporation*

Powder metallurgical techniques lend themselves very readily to fibre incorporation. Powders and discontinuous fibres are blended using standard powder mixing equipment. The resulting mixtures are then pressed and sintered or hot pressed. Modern developments in powder fabrication such as sinter forging are well suited to composite materials manufacture. Billets produced in this way, or by liquid state techniques, are formed into usable shapes by conventional rolling, swaging and extrusion processes. These latter processes perform in addition the essential function of aligning the fibres. They are also employed to bond together clad fibres produced by coating continuous fibres by drawing through a melt, by electrodeposition or by vapour deposition. During these metal working processes care must be taken to avoid extensive fibre damage. Shear stresses transferred to the fibres during extensive deformation will cause fibre breakages which in the limit can result in no fibres being longer than the critical length. However, the critical length under the deformation conditions of high temperature, at which matrix shear strength is low, is extremely high, so that the critical length should be easily exceeded at the lower service temperature of the composite. Composite structures have also been manufactured by hot rolling or diffusion bonding of a sandwich structure in which continuous wires are laid between metal sheets.

The direct manufacture of fibre composite materials using metallurgical techniques to produce a fibrous morphology of a reinforcing phase has been studied. Growth of fibrous phases by solid state transformation is practicable from the point of view of reinforcement in alloys showing a eutectoid transformation. A banded structure is produced. Some success has been achieved in the ferrous field, which has the advantage of a martensitic transformation to strengthen the fibrous phase. In terms of modulus improvement, however, this approach offers no promise and an alloy composed of cementite bands in a ferrite matrix would have more potential. A similar direct route involves the alignment of a strong fibrous or lamellar phase by deformation of the alloy at elevated temperature.

(b) Liquid State Incorporation

The simplest method of liquid state incorporation is to pour molten metal around pre-arranged fibres. This approach has been used with some success but is unlikely to yield high volume or high volume fraction composites. Problems of void formation occur at high volume fractions. Vacuum infiltration has been widely employed to alleviate voidage but fibre packing problems remain. For discontinuous fibres, hot pressing of a powder/fibre blend at a temperature such that the final stages of pressing are carried out in the presence of a liquid phase has met with considerable success. Where the fibres are very resistant to the molten metal and can be dispersed in it, pressure diecasting becomes a possibility. Aluminium alloys reinforced with silicon carbide whiskers have been manufactured by this route, which would seem to hold significant technological promise.

The direct manufacture of fibre-reinforced composites by a solidification route is possible. By analogy with the eutectoid transformation in the solid state, the eutectic growth of a fibrous morphology can be performed. Unidirectional solidification of an alloy of eutectic composition produces a fibrous or lamellar structure aligned along the axis of solidification. However, severe problems exist and it would seem that technological exploitation is unlikely. The growth of intermetallic particles during solidification can be expected to yield more success in the long term. These phases are strong and stiff and they can be induced to grow in fibrous form in the direction of highest heat extraction. Fig. 24 shows a technological aluminium-silicon alloy reinforced with intermetallic particles. The volume fraction of intermetallic phase can be varied according to requirements. After solidification it is possible to increase the volume fraction of reinforcement by raising the temperature to just above the solidus and pouring off the liquid formed.

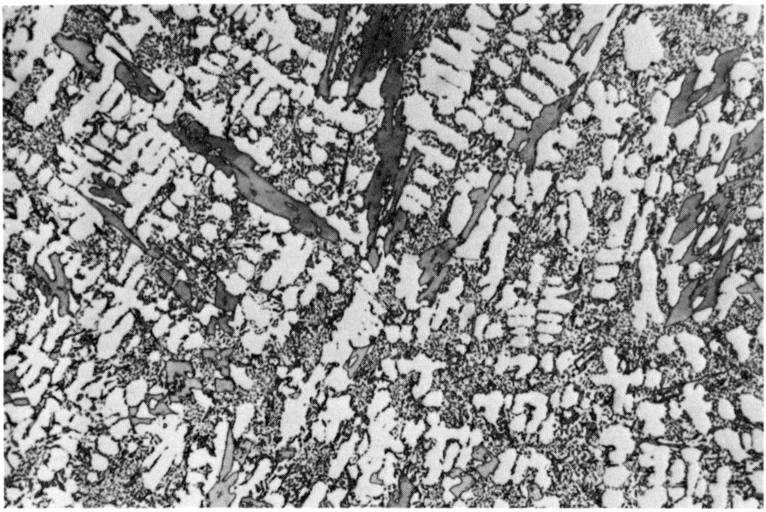

Fig. 24. Aluminium–silicon alloy reinforced with Al_3Ni intermetallic particles $\times 250$.

Thermal spraying has been used to produce fibre-reinforced prismatic shapes. A mandrel of the required

Fig. 25. Winding equipment used in the manufacture of fibre reinforced composites by plasma spraying.

form is manufactured and a layer of flame or plasma sprayed metal applied above a release agent. A layer of continuous reinforcement is then applied using equipment of the type shown in Fig. 25, and a second deposit of metal is sprayed on. The process is repeated until the required number of layers has been achieved. The composite sleeve is released from the mandrel, cut up into segments and hot-pressed to promote matrix-fibre bonding and to densify the material. A section of an aluminium/stainless steel wire composite produced in this way is shown in Fig. 26.

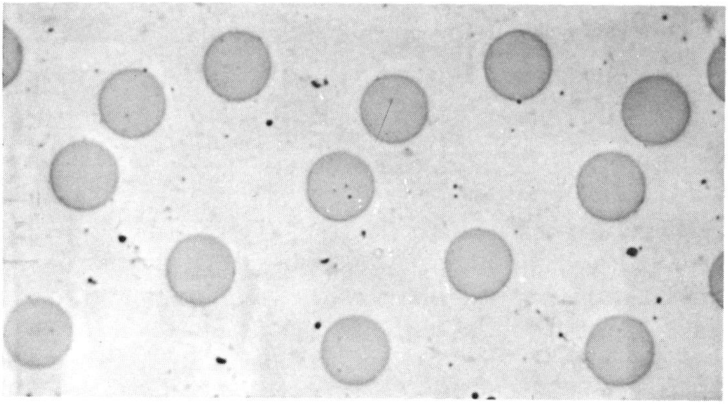

Fig. 26. Section of aluminium/stainless steel wire composite produced by plasma spraying and hot pressing.

(*c*) *Atomic State Incorporation*

Electrodeposition is the best known example of a process performed on the atomic scale. The coating of a shaped anode with a metal by electrodeposition during which a continuous reinforcement is wound on the anode would seem to be a practicable process. Less practicable, apparently, is the combination of electrodeposition of the matrix with electrophoretic deposition of a discontinuous fibrous phase.

Electroless deposition has been studied for composite manufacture, but it would appear that this process is technologically limited to the provision of protective or bonding layers on non-electrically conductive fibres.

Vapour deposition is a similar atomic scale process in which metal vapour is deposited on fibres. Chemical vapour deposition, similar to the process used in the production of boron fibres, involves decomposition of a gaseous compound on a heated substrate. This has advantages over physical deposition in which deposition occurs at all points of the treatment vessel, resulting in a very inefficient process.

Atomic-scale processes are likely to find applications only where other techniques of composite manufacture are impracticable. They are more likely to find wider application in the intermediate processes of fibre manufacture and fibre coating.

CERAMIC MATRIX COMPOSITES

Fibre-reinforced ceramics can be prepared by hot-pressing a mixture of ceramic powder and reinforcing fibre. If metal wires are used with a coefficient of thermal expansion higher than that of the ceramic, compressive stresses are produced in the matrix on cooling and this results in improved mechanical properties. Hot pressing tends to align fibres in random array perpendicular to the pressing direction.

The casting of a molten ceramic around pre-stressed continuous refractory metal wires produces a matrix compressive stress. A melt approach must suffer from the limitations discussed above in relation to metal composites, however. Substantial benefits may derive from low volume fractions of reinforcements in the ceramic field, so the problems may not be so acute.

Slip casting is a process involving the casting of a slurry of ceramic, known as the slip, into a porous absorbent mould. The carrier liquid is absorbed by the mould and a weak but handlable shape is produced. Fibres can be dispersed in the slip to be deposited in the ceramic body in random array, or a mat of fibres can be placed in the mould and infiltrated with the slip. The green body is dried and fired after removal from the mould.

Fig. 27. Plasma spraying—used in the manufacture of ceramic and metal composites.

Ceramic coatings and hollow bodies can be prepared with high density by plasma spraying (Fig. 27). The reinforcement of such material by a process similar to that described for metal matrix composites is feasible. Small diameter wires are used to obviate problems of shadowing (Fig. 28) and the need for hot pressing. Pretensioning of the wire is performed.

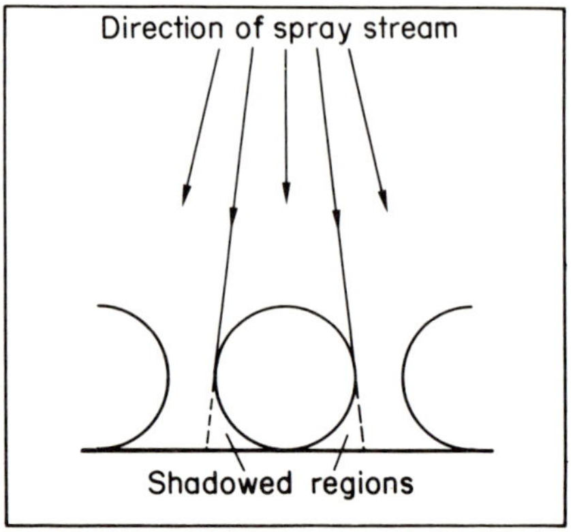

Fig. 28. Schematic illustration of shadowing occurring during the plasma spray manufacture of a fibre-reinforced composite.

5. Properties of Fibre-Reinforced Materials

PLASTIC MATRIX COMPOSITES

The properties of a range of plastic matrix composites are represented in Table 7.

Theory predicts that composite strength and modulus should increase with increase in volume fraction of fibre. At high fibre contents, however, maximum values are reached above which strength and modulus fall. This is ascribed to the difficulty of dispersing the fibres throughout the resin at high volume fractions so that voids result. The volume fraction at which the maximum values occur is dependent on the manufacturing process and the form of the reinforcement. A further practical problem is cure shrinkage, and this might be expected to cause departures from theory, but in practice no correlation between this factor and mechanical properties has been established.

The tensile properties of unidirectional fibre-reinforced plastics are found to be only marginally influenced by matrix properties. This is to be expected since in a well-designed composite the fibres bear the majority of the load. The low modulus values of plastic matrices result in effectively the whole of the load being borne by the fibres. Similarly, the ultimate elongation of the plastic has minimal effect on tensile properties, if it exceeds that of the reinforcement.

The flexural properties of parallel fibre-reinforced plastics are influenced by matrix modulus as shown in Fig. 29. Other similar flexural property/matrix strength correlations appear to hold. Fig. 29 exaggerates the practical importance of this effect, however, since most resins have modulus values between 3 and 4 GN/m^2, in which range composite property changes are only marginal.

Table 7. Properties of a range of fibre-reinforced plastics

SYSTEM: MATRIX	SYSTEM: VOLUME FRACTION OF FIBRE (× 100)	RELATIVE COST (**)	YOUNG'S MODULUS (GN/m^2)	ULTIMATE TENSILE STRENGTH (MN/m^2)	SPECIFIC GRAVITY	REMARKS
Epoxy	14 S-glass	2	12	520	1·38	Unidirectional continuous
Epoxy	60 S-glass	2	51	2000	1·95	Unidirectional continuous
Epoxy	60 B-fibre	100	140	920	2·06	Unidirectional continuous
Epoxy	73 B wire	350	230	1000	1·69	Unidirectional continuous
Epoxy	50 C-fibre	10	190	590	1·47	Unidirectional continuous
Epoxy	35 Si_3N_4 whisker	5	100	310	1·90	Unidirectional
Epoxy	40 Al_2O_3 whisker	100	166	500	1·74	Unidirectional
Nylon	40 Chopped glass	2	11	200	1·41	Elongation 5% unidirectional
Polycarbonate	40 Chopped glass	2	10	130	1·44	Elongation 5% unidirectional
Polystyrene	30 Chopped glass	2	8	97	1·28	Elongation 1% unidirectional
Polyester	22 Chopped glass	2	7	140	1·55	Isotropic in plane of reinforcement
Epoxy	60 S-glass fabric	2	22*	660*	1·95	Bidirectional
Epoxy	60 E-glass fabric	2	21*	370*	1·95	Bidirectional
Polyester	40 E-glass fabric	2	20*	320*	1·66	Bidirectional

* Average values: warp values exceed fill values by up to 10% ** Steel wire — 1 per unit of weight

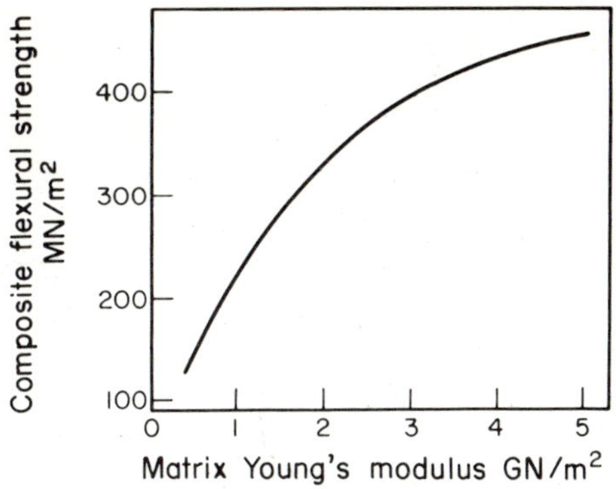

Fig. 29. Relationships between composite flexural strength and matrix Young's modulus.

Interlaminar shear strength is a property of some importance in polymer matrix composites since, like transverse strength, it tends to be very low in relation to tensile properties. Both are controlled by the weaker of matrix or interface, and both of these are weak in plastic matrix composites. Interlaminar shear strength falls with increase in fibre content, indicating either that the interface is the limiting factor or that voidage problems increase as fibre content rises. Similarly, transverse strength can be expected to follow this trend. There is some evidence that matrix elongation may affect transverse strength especially in the presence of moisture. Microcracking can occur near fibre ends on loading at low stresses transverse to the fibres. If a high elongation matrix is used, such premature microcracking can be obviated, with consequent strength improvement. Interlaminar shear strength is dependent on matrix tensile strength, but it is not clear whether a matrix effect or an interface effect is responsible for the dependence.

The compressive strength of plastic composites falls with decreased resistance to fibre buckling. This indicates that compressive strength increases with increasing matrix modulus and yield strength. Interlaminar shear strength is also directly related to compressive strength, since ultimate failure tends to occur as delamination. Other factors such as tensile properties and elongation are related to compressive strength, but it is considered that these are merely indirect effects.

The importance of interfaces in the properties of fibre-reinforced plastics has already been stressed. In this respect in the glass fibre field especially, coupling agents play a major role. These are organometallic or organosilane finishes which are applied to the glass fibre and have the property of interacting with both fibre and polymer. Coupling agents increase both dry strength (by up to 40%) and strength after exposure to moisture (by up to 200%). Their mode of operation is uncertain but the property improvements are significant. Little experience in this approach has been gained outside the glass fibre field. The interfacial mode of crack interruption is operative in fibre-reinforced plastics and accounts for the high toughness of these materials.

Voids exercise a marked influence on composite interlaminar shear strength and compressive strength. Tensile properties are not seriously affected, however, since composite tensile properties are only slightly dependent on matrix properties. Flexural properties can be expected to be adversely affected by voids since they will exert a harmful effect in the compressive regions.

The fatigue properties of fibre-reinforced plastics are confounded by problems of heat dissipation arising from high internal friction and low thermal conductivity. However, epoxy matrix composites perform well under fatigue conditions and this has been attributed to the strength of the fibre-matrix bond. Otherwise it appears that matrix properties do not exert a major effect. Optimisation of composite tensile strength does not necessarily imply optimisation of fatigue strength since the failure mechanisms are quite different. Fatigue strength increases with increase in modulus of the reinforcement.

METAL MATRIX COMPOSITES

The properties of a range of metal matrix composites are presented in Table 8.

SYSTEM						
MATRIX	VOLUME FRACTION OF FIBRE (× 100)	RELATIVE COST (*)	YOUNG'S MODULUS (GN/m^2)	ULTIMATE TENSILE STRENGTH (MN/m^2)	SPECIFIC GRAVITY	REMARKS
Aluminium	25 Steel wire	10	100	1200	3·6	
Aluminium	40 Be wire	200	160	550	2·3	
Aluminium	50 SiO_2	10	71	900	2·4	
Aluminium	50 B-fibre	100	225	760	2·7	
Aluminium	35 Al_2O_3 whisker	100	180	1100	3·1	
Aluminium	10 $NiAl_3$	10	80	380	2·8	Unidirectionally solidified
Aluminium	50 $CuAl_2$	10	130	270	3·4	Unidirectionally solidified
Al 10 Si	15 Al_2O_3 whisker	50	80	280	2·9	Random orientation
Silver	24 Al_2O_3 whisker	100	110	1600	8·5	
Copper	77 Tungsten wire	50	300	1800	16	
Titanium	20 Molybdenum wire	50	150	660	5·3	
Cobalt	30 Tungsten wire	50	260	740	10·7	
Nickel	75 Boron fibre	200	350	2600	4·0	
Iron	36 Al_2O_3 whisker	100	280	1600	6·7	
Stainless steel	18 Tungsten wire	50	230	400	10	

* Steel wire = 1 per unit of weight.

Metal matrix composites follow, more closely than the other systems, the theoretical treatments outlined in Chapter 2. Since many of the relationships were derived from and checked by work on metal-metal systems, for examples copper–tungsten wire and silver–steel wire, this is hardly surprising. However, metals do exhibit many characteristics required for successful fibre reinforcement in that they are reasonably strong yet ductile and they bond well to a wide range of fibres.

The stress–strain behaviour of a typical fibre-reinforced metal is shown in Fig. 8 for both ductile and brittle fibres. Deviations from the predicted tensile strength values can arise due to inequalities in fibre strength, poor alignment, discontinuity, low strength interfaces or non-uniform distribution. Degradation of fibre strength during processing can also result in lower strength being achieved.

With brittle fibres the strain at failure is approximately equal to that of the fibres. Ductile fibres when tested individually fail by necking. On incorporation in a ductile matrix hydrostatic constraint arises to prevent necking with the result that higher elongations are achieved. In a fibre which work hardens this results in a positive deviation from the predicted strength.

Fibre-reinforced metals exhibit the predicted elastic behaviour up to the yield strain of the matrix, after which the slope of the stress–strain curve alters, being governed at a given composite strain by the slope of the matrix stress–strain curve as predicted in Chapter 2. The yield stress and the ultimate stress of a composite containing ductile fibres follow a rule of mixtures relationship.

From a practical point of view the most important part of the stress–strain curve is that concerned with elastic behaviour of the fibres. For much of this region a metal matrix may be straining plastically. This results in a possibly beneficial effect which can be brought about by prestraining a part subject to fatigue stresses. On unloading a composite in which the matrix has yielded the matrix stress falls to zero while the fibres are still under a tensile load. The stress in the fibre reduces further until it is balanced by the compressive stress produced in the matrix. Under these conditions initiation and propagation of fatigue cracks is rendered more difficult. Fig. 30 illustrates this effect.

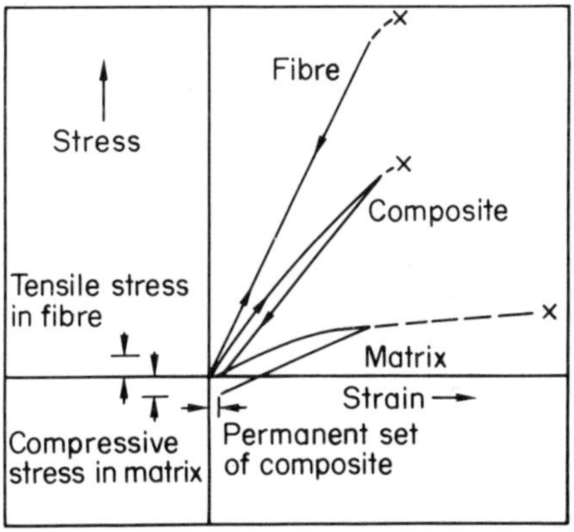

Fig. 30. The effect of prestrain on the residual stress levels in a fibre-reinforced composite material.

The concepts of V_{min}, below which failure of the matrix does not result in failure of the composite, and V_{crit}, below which no reinforcement results, are found to exist in practice. The values of V_{min} and V_{crit} increase with decrease in the difference between matrix and fibre strengths.

When a composite is reinforced with discontinuous fibres, strengths are in general less than those predicted by rule of mixture considerations. It is found that there is more scatter in the strengths produced by discontinuous fibre-reinforced metals, and this is most probably attributable to the lower degree of alignment achieved with discontinuous fibres. The shear strength of the fibre matrix interface is more important when discontinuous fibres are used. For strength to approach that predicted by the rule of mixtures it is necessary to exceed the critical length or critical aspect ratio by a large factor dependent on the nature of the matrix and of the fibre. When the fibres are close to the critical length they tend not to fracture but to pull out of the matrix.

The effect of fibre alignment is extremely critical. The strength of composites falls drastically when the stress axis is only a few degrees away from the fibre axis in a similar fashion to that shown in Fig. 11. By analogy with this any fibres in a composite misaligned by more than a few degrees will provide little, if any, reinforcement. Randomly orientated fibres provide reinforcement but at a much lower level than that provided by unidirectional reinforcement. In fact it seems from the few results available that randomisation in two dimensions provides one third of the unidirectional strength and randomisation in three dimensions provides one sixth, as predicted by theory.

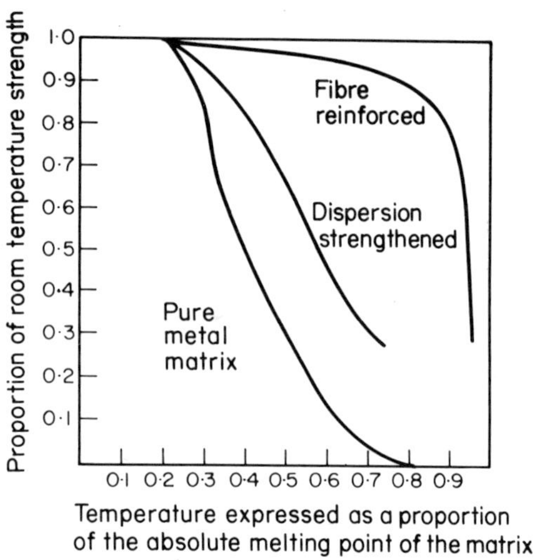

Fig. 31. Comparison of fibre reinforcement and dispersion strengthening in their effect on the elevated temperature strength of a metal matrix.

The behaviour of fibre-reinforced metals under conditions of other than simple tension is not well documented. In compression, metals are strong and it seems that the compressive strength of composites will be high. It is unlikely that delamination failures will result since metal matrices are stiff and interfacial strengths are high. Metals reinforced with low volume fractions of fibres, less than 20%, exhibit much higher fatigue life as compared with unreinforced material. In general, crack propagation rates are lower for reinforced metals and the crack instability length is greater. Matrix tensile residual stresses can result in reduced fatigue strength but these can be offset by prestraining as discussed above. The shear strength of metals and metal fibre interfaces is in general much higher than that occurring in reinforced plastics, so low interlaminar shear strength values are no problem in metal matrix composites. At elevated temperatures the properties of fibre-reinforced metals are still predicted by the rule of mixtures. Creep properties are markedly improved by the presence of refractory fibres. Fig. 31 shows the powerful effect of fibre reinforcement in improving the elevated temperature strength of a ductile metal matrix.

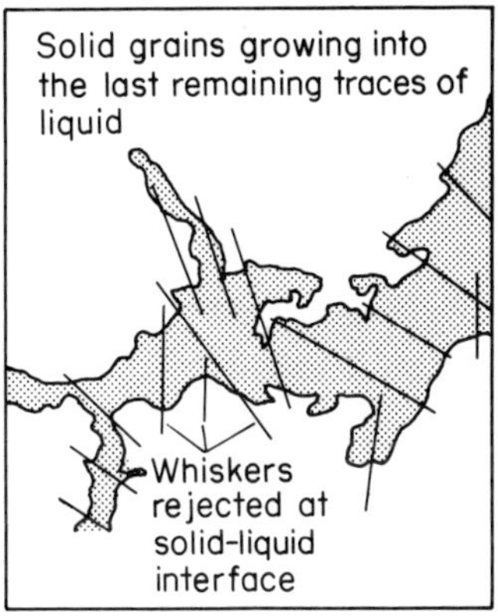

Fig. 32. Schematic illustration of the rejection of whiskers at the solid–liquid interface in a solidifying metal $\times 500$.

The properties of casting alloys have been markedly improved by the incorporation of a very low volume fraction of refractory fibres. These appear to be rejected by the solid-liquid interface and bridge the grains of the alloy to provide a network of high strength and rigidity as shown schematically in Fig. 32.

CERAMIC MATRIX COMPOSITES

Ceramics tend to have high modulus values and exhibit no sensible plastic flow. It is difficult therefore to reinforce them in the conventional manner since transfer of stress to a fibrous reinforcement is only marginal under normal conditions. Low modulus ceramics such as mullite or whiteware and glasses have shown marked increases in tensile properties and modulus of rupture due to the presence of metal fibres. Thermal shock properties are also markedly improved. If a fibre is incorporated at elevated temperature in a ceramic of high coefficient of thermal expansion, then the matrix will be stressed in tension on cooling, microcracking can occur and strength reductions can result. To overcome the problems of reinforcement of high modulus ceramics it has been found beneficial to prestress the fibres to build in a tensile stress. Fibre reinforced ceramics prepared in this way show markedly improved mechanical properties. Ceramics have also been successfully reinforced with wire mesh.

In general, successful fibre reinforcement of ceramics results in a marked reduction in the brittleness of the materials which manifests itself in improved toughness and thermal shock behaviour. The latter is particularly important since the temperature capability of ceramics is well in excess of that of metal matrices. Data on the properties of fibre-reinforced ceramics are relatively scarce and a consistent quantitative picture cannot be provided at this time.

6. Applications and Outlook

Man-made fibre-reinforced composites have been in use in many well known areas—for example, fibreglass car bodies, brake linings—for many years. It is not the purpose here to survey such uses, however, but to concentrate on more sophisticated, advanced applications. In view of the recent and rapid development of this field any review can be of only transient validity. No attempt will be made to be exhaustive. The examples discussed, even when obsolete, will serve to indicate the design and applications philosophy in fibre reinforcement and it is in this that any lasting value of this section lies.

APPLICATIONS

The major activity in advanced composites has been associated with the aerospace industries. Not only are totally new materials requirements continually arising in these fields but large budgets are available to develop new materials. This is a reflection of the costs involved in lifting each unit of weight off the ground. In the extreme case the cost of lifting 1 kg out of the atmosphere is of the order of £10,000 ($25,000). Thus high strength, lightweight materials are at a premium in aerospace and these are the major attributes of advanced fibre reinforced composites.

The most widely publicised application of a fibre-reinforced composite in the aerospace field serves to indicate the problems associated with the introduction of a new generation of materials. The carbon-fibre reinforced plastic (CFRP) compressor fan blades manufactured by Rolls Royce for the RB211 advanced technology engine received as much publicity in the conceptual stage—being the first major use of carbon fibre—as they did in the disastrous application stage. Problems of water droplet erosion and the dangers of birdstrike ruled out these blades for use in the Lockheed Tristar, and titanium blades, developed in a back-up programme at Lockheed's insistence, were adopted instead. The immediate consequences of the failure of the CFRP blades, although not directly occasioned by this, may serve a cautionary purpose to those proposing to embark upon a costly development in the composites field. However, the problems have been overcome by the incorporation of metal inserts to improve erosion resistance and ductile metal foils to diffuse high intensity stresses. The solution of the problems of this composite material was to produce an even more complex composite. The future for the application is brighter than in the past since it is proposed to use the composite in later versions of the RB211. The use of glass fibre-epoxy and boron fibre-epoxy composites for similar applications is proposed in the U.S.A.

Other aircraft uses of fibre-reinforced composites, both actual and projected, include helicopter tail rotor blades in boron fibre-epoxy over aluminium honeycomb. This combination has also been suggested for industrial fans and compressors. The use of CFRP floor panels in VC 10s and Boeing 747s has been pioneered by B.O.A.C. In the F111 supersonic fighter bomber the tail is designed with boron fibre-epoxy face sheets over an aluminium honeycomb core. Boron fibre-epoxy is also used in the trailing edge panel and the wing pivot fittings. It is proposed to use CFRP for the wing tips of the F5A aircraft. Glass fibre reinforced plastic plasma sprayed with alumina has been developed for use in erosion resistant radomes. With the growing tendency towards containerisation in transport it is likely that aircraft containers of the future will be manufactured in a fibre reinforced composite material. Fibre reinforcement of turbine blades is an exciting possibility in the metal matrix field. More distant perhaps is the use of hollow filaments to serve the dual purpose of cooling and reinforcement.

Rocket motor casings in CFRP have been widely manu-

Fig. 33. CFRP rocket motor casing. (Courtesy T. A. Trigg, Bristol Aerojet Limited.)

Fig. 34. CFRP satellite structure. (Courtesy I.C.I. Limited, Summerfield Research Station.)

factured (Fig. 33). Boron fibre-epoxy has been used for the mid-section of a cylindrical re-entry vehicle. A satellite structure in which the principal material was CFRP has been manufactured (Fig. 34). Carbon fibre cloth impregnated with a phenolic resin has shown promise for ablative nose cone applications. Much interest has been associated with carbon–carbon composites in which a carbonaceous matrix is reinforced with carbon–graphite fibres. An aircraft brake for use in Concorde has already appeared and other applications include heat shields, leading edges and rocket nozzles.

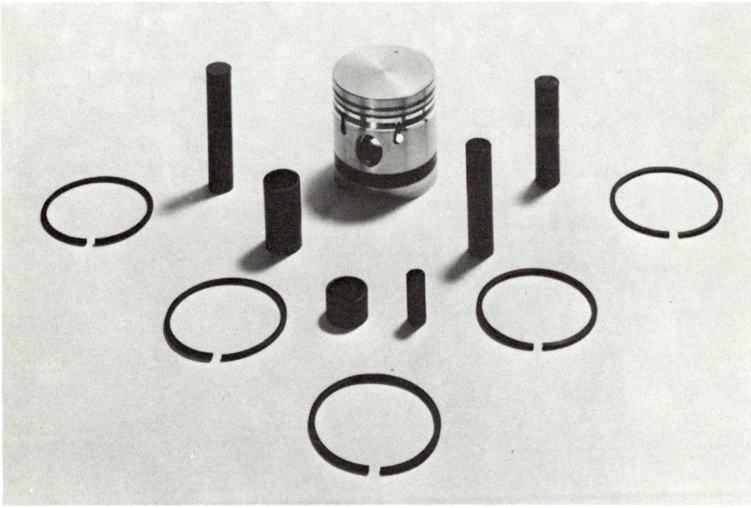

Fig. 35. A selection of piston rings, gudgeon pins and a piston incorporating the use of CFRP.

In reciprocating machinery, to come down to earth, much energy is wasted due to the inertia of the reciprocating parts. In advanced equipment very high speeds are achieved and the problem is particularly severe. The use of a CFRP heddle frame in a textile machine allowed even higher speeds to be attained with additional operating benefits. The selection of automotive components shown in Fig. 35 was developed for this purpose so that bearing loads could be reduced. The weight of the CFRP gudgeon pin is one third that of the equivalent steel product. The piston rings are light in weight and possess low coefficient of friction. The pin has operated in an engine but has not been as successful as the rings. The CFRP belt around the piston skirt is there primarily for thermal control, the expansion coefficient of carbon fibre being approximately zero. The piston shown in Fig. 5 suffered from a fatigue problem in the region of the insert. It was not permissible to increase the relevant section thickness since this would increase the reciprocating mass. Fibre reinforcement was used so that improved fatigue performance could be achieved without a weight penalty. The high quality of the joint between reinforced insert and piston casting is shown in Fig. 36.

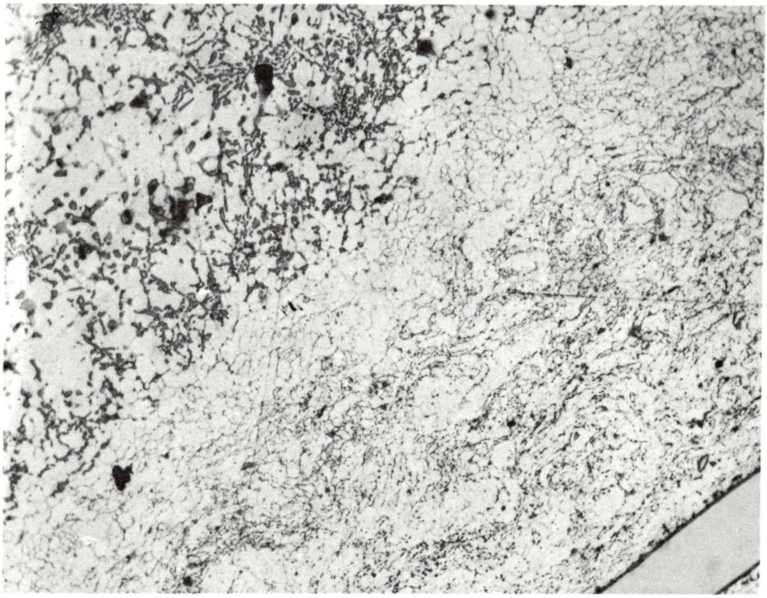

Fig. 36. Interface between fibre-reinforced insert and cast piston body ×100.

The low coefficient of friction of carbon fibres perpendicular to the fibre axis has already been referred to, and this offers incidental advantages in many applications. In some designs however this is a major feature. Sliding vanes for unlubricated compressors and exhausters are manufactured in CFRP where the self-lubrication is as important as the high strength. Asbestos reinforced plastics are used in lubricated compressors (Fig. 37). The manufacture of gears in CFRP is a similar dual function application, as were the piston rings and gudgeon pins discussed above. Bearings fail mechanically by fatigue and the benefits of fibre reinforcement are obvious. This use of carbon fibre again allows of the dual function approach.

Fig. 37. Section of a rotary compressor using asbestos reinforced rotor blades. (*Courtesy TAC Construction Materials Limited.*)

In civil and chemical engineering the use of fibre composites is increasing. Concrete reinforced with steel rods has been used for many years, but there is a trend towards the use of steel wire in cement and plaster. To the handyman who has known his plaster crack on drying, the benefits of fibre reinforcement are immediately apparent when he uses nylon or glass reinforced plaster. Steel wire reinforced Portland cement and refractory castables show substantial strength benefits when containing only a few per cent by volume of randomly orientated short wires. The use of flexibilised reinforced plastics of high impact strength is projected for building use. Rubbery particles precipitate in the glassy matrix at gelation to produce a resilient matrix which is reinforced with glass fibres for high strength. The use of glass-reinforced plastics in civil engineering for vessels, valves and pipe-work is well established. Alkaline attack is a problem, however, and it has been suggested that improved performance could result from the use of CFRP.

Fig. 38. CFRP squash rackets. (*Courtesy W. Paton, N.E.L.*)

From the many thousands of present and projected uses of fibre-reinforced composites the following selection is made. The use of CFRP in submersible vessel shells and in the joint U.K., Netherlands and West German gas centrifuge is projected. Artificial prostheses need to possess high strength, high lubricity at least initially, and compatibility with human tissue. It is suggested that a carbon fibre-reinforced plastic approaches these requirements more closely than any other material. In the sports goods field CFRP has already been successfully applied to skis, water skis, squash rackets (Fig. 38), golf clubs (Fig. 39), cricket bats and dinghy masts. The esoteric benefits of carbon fibres in sports goods, coupled with the seemingly bottomless pocket of the public in its pursuit of leisure, would seem to ensure a large and growing market for carbon fibres in this field. Tyres have been reinforced with nylon and with steel, each with its own particular advantages, for many years. Glass fibre is now coming on the scene and this may combine some of the benefits of nylon and steel to give warning of breakaway coupled with long life. Carbon fibre is projected for this application too, although the benefits are not particularly apparent.

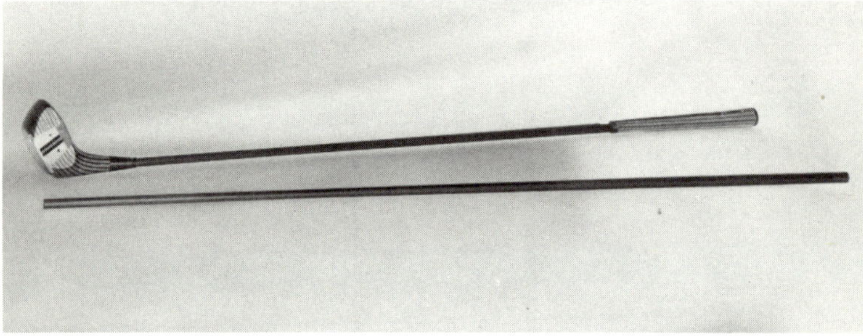

Fig. 39. CFRP golf clubs. (*Courtesy W. Paton, N.E.L.*)

Electrically conducting plastics usually contain metal powder or flake. Modern developments in this area include the incorporation of silver-, copper-, or nickel-coated glass fibre, steel wire and carbon fibre. The scale of modern composite technology is well demonstrated by the manufacture of the Royal Navy minesweeper, H.M.S. *Wilton* in glass-reinforced polyester. The total weight of the moulding was in the region of 200,000 kg. The largest metal matrix composites to date are reputed to be 2·5 m × 0·3 m × 16 mm panels of aluminium reinforced with steel wire for cryogenic applications. Casting alloys reinforced with low volume fractions of refractory fibres as discussed above are an interesting possibility. The feasibility of the production of silicon carbide fibre-reinforced silicon nitride, in which incidentally the fibres are naturally in tension and the matrix in compression, has been demonstrated and promising mechanical properties result. Fibre-reinforced ceramics are suggested for radomes, printed circuits, thermal insulators, extrusion dies and nuclear engineering uses. Experimental rocket nozzle inserts in slip-cast tungsten wire-reinforced titanium carbide have been successfully tested. Reinforced ceramic coatings are suggested as linings for ramjet combustion chambers and plasma chambers. A stainless steel wire reinforced zirconia coating performed successfully in a heat exchanger and it was concluded that in general for successful application the matrix should be in compression at temperature.

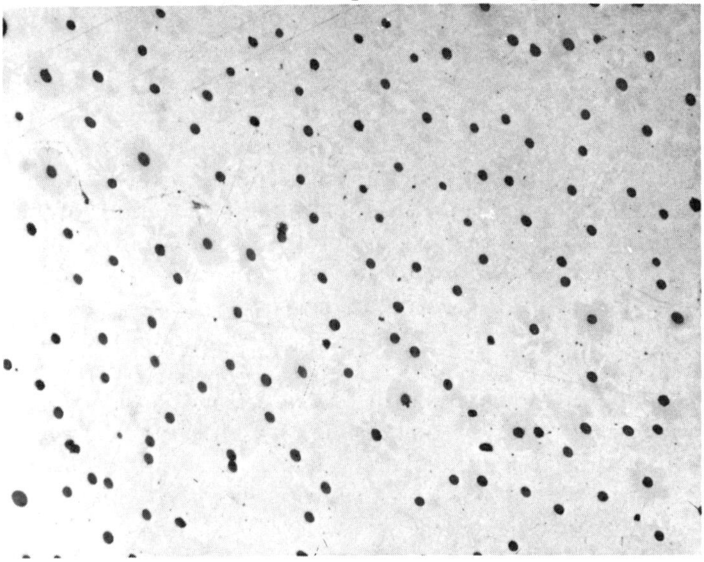

Fig. 40. Carbon fibres protected with an electroless nickel deposit incorporated in a zinc matrix.

In this survey of applications most emphasis has been placed on plastic matrix composites, usually reinforced with carbon or boron fibres. Exploitation of metal matrix composites will follow with satisfactory solution of the compatibility problem (Fig. 40). Technologically plastics are the easiest matrices to exploit in this way and politically these two fibres carry much weight in their respective countries of origin. This emphasis should not blind the designer to the benefits of other reinforcements. Usually reinforcements will be chosen primarily on the basis of stiffness. Each use will be different, however, and it is not possible to lay down rules. In making a choice of material, it is necessary to survey the range of those available, keeping in mind the detailed requirement and the fabrication route.

OUTLOOK

Although projected applications of fibre-reinforced composites abound, successful mass exploitation is not likely to be with us for some time. Industry is at present geared to producing monolithic blocks of material from which components are manufactured. Composites must be produced as recognisable components in one process and the stockholding approach is impracticable. This requirement for a radically different materials policy is one of the causes of the, to many, prohibitive cost of composites. A new approach will result in more realistic cost levels. Reinforcements are, in the main, very costly, but few are inherently costly. The carbon fibre syndrome indicates that high volume, low cost fibres do not arise overnight. Eventually, however, with increasing market demand, fibre costs will become more reasonable. Low cost fabrication routes involving plasma spraying, diffusion bonding or casting and the fibre-reinforced insert approach can be expected to be developed further in preference to the more costly processes.

Thus there is no major insurmountable economic barrier. But what of composite properties? Toughness problems remain in the highest strength composites. It has been indicated that the interfacial mode of crack stopping may be ineffectual. J. G. Morley has proposed a fascinating solution to the toughness problem in which a duplex fibre is used which has a weak internal interface (Fig. 41). High transverse strength is achieved by a strong bond between the external interface and the matrix. Toughness is derived from the weak internal interface. Silicon carbide coated carbon fibre is proposed as a suitable system. The external bond is further improved by overgrowths of whiskers on the surface of the fibre to provide additional mechanical keying. The precipitation of rubbery particles in glassy polymers has been mentioned. This results in toughness at the expense of stiffness. Such a trade-off is in fact employed by nature. Wood, teeth and bone possess only a small fraction of the strength they could theoretically possess, but they are extremely tough. In these days when fracture mechanics and flaw sizes loom ever larger, high strength materials with barely detectable critical crack lengths—the lengths at which a crack will grow in an unstable manner—are seen to be a liability. It may well be that the future of fibre-reinforced materials lies in medium strength, tough materials.

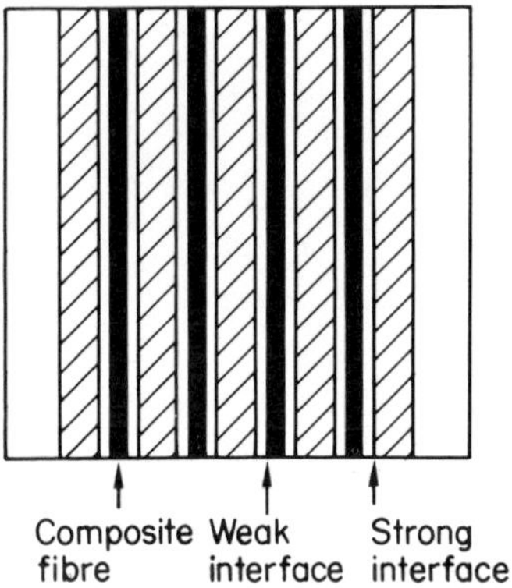

Fig. 41. Proposed mechanism for achieving high transverse strength and high toughness in a fibre composite material.

It is possible to envisage a future in which composites are widely used with their properties tailored to suit particular applications. This will require, however, an ability on the designer's part to understand his materials and to accept anisotropy as the price of good unidirectional or bidirectional properties. Reinforcement in three dimensions for strength as opposed to toughness is not considered likely. New joining techniques will need to be established. The future for fibre-reinforced composite materials looks promising, although it is by no means a marketing man's dream. Progress is likely to be slow but steady. Spectacular advances can result in spectacular retreats.

Bibliography

1. "A discussion on new materials."
Proceedings of the Royal Society, **282**, A (1964), Royal Society, London.
A good idea of some reasonably early, but visionary thoughts on the subject is provided in the following papers: "Strong solids", A. H. Cottrell, p. 2. "Some consideration in the design of engineering materials based on brittle solids", J. E. Gordon, p. 16. "Strong fibres and fibre reinforced metals", J. G. Morley, p. 43.
2. "A discussion on strong fibrous solids."
Proceedings of the Royal Society, **319**, A (1970), Royal Society, London.
A thorough updating of reference 1.
3. *Metallurgical Reviews*, **10** (1965), Institute of Metals, London.
Contains the two definitive works: "The principles of the fibre reinforcement of metals", A. Kelly and G. J. Davies. "Experimental aspects of fibre reinforced metals", D. Cratchley.
4. "Modern Composite Materials", ed. L. J. Broutman and R. H. Krock, Addison-Wesley, Reading, Mass., U.S.A. (1967).
Reviews the whole of the composites field, with good coverage of the plastics area.
5. "Ceramic fibres and fibrous composite materials, H. W. Rauch, Sr., W. H. Sutton and L. R. McCreight, Academic Press, New York and London (1968).
An in-depth survey of the state of the art, c. 1967.
6. "Fibre composite materials", American Society for Metals, Ohio (1964).
An early review of the subject which provides a rare survey of the ceramic field.
7. "Fibre reinforced materials", G. S. Holister and C. Thomas, Elsevier Publishing Company, London (1966).
A basically analytical survey of fibre reinforcement.

INDEX